Mapping Our World
Using GIS

Anita M. Palmer

Roger Palmer

Lyn Malone

Christine L. Voigt

Student Workbook

ESRI PRESS
REDLANDS, CALIFORNIA

Contents

Module 1, Lesson 1

The basics of Arcmap

- Activity
- Answer sheet

Module 1, Lesson 1

The basics of ArcMap

ArcGIS is made up of two programs: ArcMap and ArcCatalog. This computer activity will show you how to start the ArcMap program. You will be guided through the basics of using ArcMap to explore maps. After you do this activity, you will be prepared to complete other GIS activities.

Step 1:   Open a map document

1.   Double-click the ArcMap icon on your computer's desktop. (If you do not have the icon on your desktop, click Start, All Programs, ArcGIS, and ArcMap.)

2.   When the ArcMap start-up dialog box appears, click **An existing map**.

3.   Click the OK button.

In ArcGIS, maps are saved in files called map documents. A map document has been created for you to use in this exercise.

4. Navigate to the module 1 folder (**OurWorld2\Mod1**) by clicking on the down arrow for the "Look in" drop-down menu. Then click Local Disk (C:), the OurWorld2 folder, and the Mod1 folder.

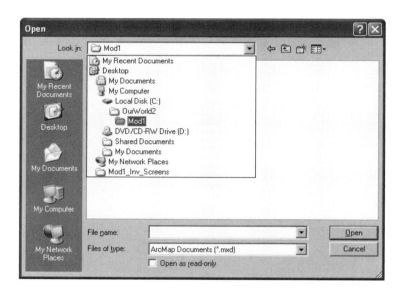

First make sure the correct drive is selected in the "Look in" drop-down menu. If the disk drive you need is not listed, click the My Computer or My Network Places icon on the left side of the dialog box to access the disk you want. Then double-click folders to open them.

5. Choose **Module1.mxd** from the list.

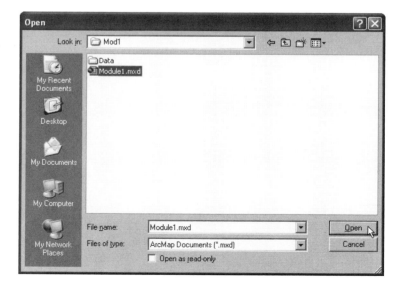

ArcGIS map document files end with the three-letter extension .mxd. If file extensions are turned off (hidden) on your computer, you won't see .mxd, and you should choose **Module1** from the list. Having file extensions turned on (visible) is not required to complete this activity.

6. Click Open.

When the map document opens, you see a map of the world. The ArcMap window's title in the top left corner of the window contains three pieces of information: the name of the map document (Module1.mxd), the program (ArcMap), and the level of the program (ArcView). (If file extensions are not visible on your computer, the name of the map document will be Module1.)

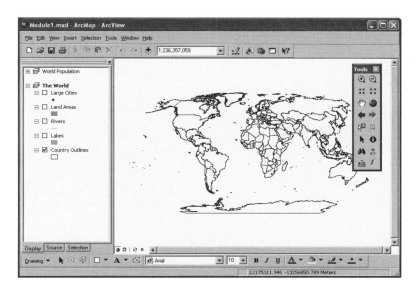

7. Look at the ArcMap window and observe the following:
   • On the right side you see a map.
   • On the left side you see a column that contains a list. This column is the table of contents.
   • The items in the list represent the two different maps contained in this map document: World Population and The World.
   • The items listed under The World are the five different layers of information that can be seen on this map.

8. Look at the top of the ArcMap window and notice the different menus and buttons. The menus and buttons are grouped on different toolbars. (Toolbars can be moved around, so your toolbars may not be arranged exactly as pictured on the following page.)

Title bar ➤
Main menu toolbar ➤
Standard toolbar ➤

The Standard toolbar above shows the map scale (1:256,357,059). Your map scale may be different depending on the size and shape of your ArcMap window.

9. Locate the Tools toolbar. It may be floating (not attached to the window) or docked (attached to the window).

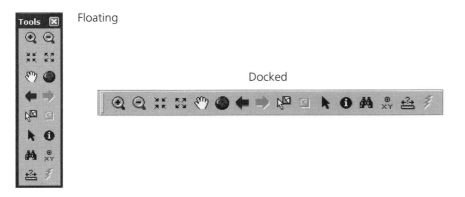

Floating

Docked

10. Experiment with the toolbar:

- If your Tools toolbar is floating, click its title bar and drag the toolbar to the gray area above the map near the other tools. The toolbar docks to the window.
- If your Tools toolbar is already docked, click the small gray bar next to the magnifying glass, and drag the toolbar off of the ArcMap window. The toolbar floats.

11. Drag your Tools toolbar to the gray line between the map and the table of contents to dock it vertically next to the map. (If it docks below the table of contents instead, drag it again until it docks where you want it.)

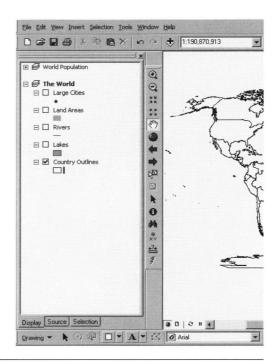

Step 2: Enlarge the ArcMap window

If the ArcMap window is small when the map document first opens, you will want to enlarge it.

1. In the upper right corner of your ArcMap window there are three buttons. Click the middle button that looks like a box.

The button you clicked is called the Maximize button. Now the ArcMap window fills your whole screen.

2. Look again at the three buttons in the upper right corner of your ArcMap window. Now the middle button looks like two boxes. (This is called the Restore Down button.) Click it. The window returns to the smaller size.

    You can also change the size of your ArcMap window by stretching it. Stretching the window instead of maximizing it allows you to organize other windows and dialog boxes as they appear.

3. Place the cursor on any corner of the ArcMap window that is not at the edge of your screen. The cursor changes to a diagonal double-headed arrow. Click and drag the window outward until the ArcMap window fills about two-thirds of the screen. Let go of the mouse button.

Step 3: Work with layers

In ArcMap, a map is made up of layers that are grouped into a data frame. Your map document has two data frames: World Population and The World.

1. Look at the table of contents and observe the following:
    *   The World data frame is shown in bold letters in the table of contents. The bold letters tell you which data frame is active.
    *   The active data frame is displayed in the map area.
    *   The World data frame contains five layers: Large Cities, Land Areas, Rivers, Lakes, and Country Outlines.
    *   Each layer has a small box in front of it. The box for Country Outlines is checked. Only checked layers are drawn on the map.

Next you will learn how to turn layers off and on and how to change their order. In ArcMap, a map is made up of layers that are grouped into a data frame. Your map document has two data frames: World Population and The World.

2. Click the check mark next to Country Outlines. The check mark goes away. The display of Country Outlines in the map area also disappears.

3. Click the box next to the Land Areas layer. The green Land Areas layer is displayed. This is called turning on a layer.

4. Click the box next to the Large Cities layer to turn it on.

5. Turn on the Rivers and Lakes layers.

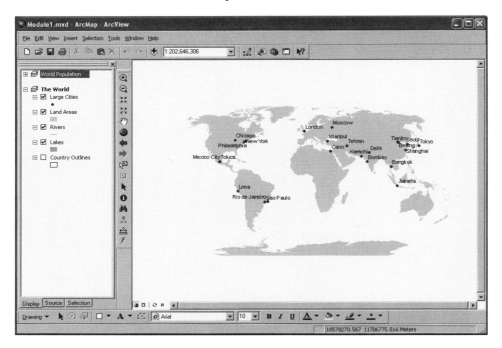

Answers to questions in this activity should be recorded on the answer sheet.

Q1 *Which layers are not visible on the map but are turned on in the table of contents?*

Next you will change the order of the layers.

6. Notice that each layer in the table of contents has a name and a symbol. In the table of contents, place your cursor on the name Lakes.

7. Click and hold the mouse button. Drag the Lakes layer up above the Land Areas layer. Let go of the mouse button.

Q2 *What happened on your map?*

8. Drag the Rivers layer above the Lakes layer.

Q3 *What happened on your map?*

Q4 *What would happen if you dragged Rivers under Land Areas?*

Whenever your maps don't appear as you think they should try the following:
a. Check to see if the layer you want is turned on.
b. Check the order of the layers in the table of contents. Layers that are represented by lines and points (streets, rivers, cities, etc.) will be covered up by layers that are represented by polygons (countries, states, etc.). You may need to drag line or point layers above the polygon layers in order to see them.

Step 4: Change the active data frame

You will now activate the World Population data frame to display the other map.

So far you have always clicked the mouse using the left mouse button. Some menus in ArcMap are accessed by clicking the right mouse button. The instructions use the words "right-click" when you need to use the right mouse button.

1. Right-click the World Population data frame title in the table of contents. On the menu that appears, click Activate (with the left mouse button). The World Population data frame becomes bold and a new map displays in the map area.

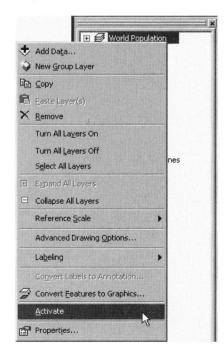

2. Click the plus sign next to World Population in the table of contents. The table of contents expands to show the layers in the World Population data frame.

Q5   *What is the name of the layer that is turned on in the World Population data frame?*

Step 5:   Widen the table of contents

Now you will widen the table of contents .

1. Move your cursor to the edge of the table of contents, in between the scroll bar and the Tools toolbar. When it is in the right place, it should look like this:

2. Click and hold the mouse button. Drag the cursor to the right until you can see the full legend descriptions. Release the mouse button. The table of contents becomes wider, and the map becomes smaller.

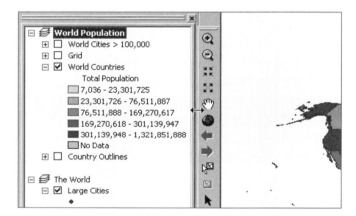

Step 6:   Obtain information about a country

The value of GIS comes from the data (information) that is attached to each map. You will see how this works by using one of the tools to access data about countries.

1. Move your cursor over the map and pause on any country (don't click). Notice that, as your cursor pauses over different countries, the name of each country is displayed. This information display is called a MapTip.

2. The Identify tool lets you see more data about your map by clicking places you are interested in. Locate the Identify tool in the Tools toolbar. Hint: To see a tool's name (ToolTip), pause your cursor over the tool without clicking.

3. When a tool is selected, it looks like the button is pushed in. If your Identify tool is not pushed in, click to select it. The Identify window appears.

M1 L1

4. Click the title bar of the Identify window and move the window so that it doesn't cover the map.

5. Move your cursor over the map without clicking. Notice it changes to an arrow with an "i" next to it.

6. Click Australia on your map. The Identify window displays information about Australia. The information you see is all the data that is available about Australia in the World Countries layer.

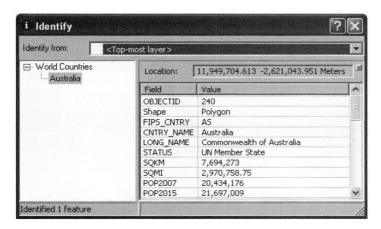

Q6 *What is the fourth listing under Field?*

Q7 *What is the fifth listing in this column?*

Q8 *What is the final listing in this column? (You will need to scroll down.)*

These words are names describing characteristics, or attributes, of Australia. Another word for attribute is field. Field names are often abbreviated.

Q9 *What do you guess the field SQMI stands for?*

Q10 *What is the number to the right of the field SQMI?*

Step 7: Compare Identify data with table data

The picture below shows part of the table that contains the data attached to the World Countries layer.

(Q11) *In the picture below, which row in the table has the attributes for Australia?*

OBJECTID *	Shape *	FIPS_CNTRY	CNTRY_NAME	LONG_NAME	STATUS	SQKM	SQMI
231	Polygon	MA	Madagascar	Republic of Madagascar	UN Member State	591712.5	228460.2
232	Polygon	MP	Mauritius	Republic of Mauritius	UN Member State	1413.31	545.68
233	Polygon	MF	Mayotte	Territorial Collectivity of Mayotte	Territorial Collectivity of France	268	103.48
234	Polygon	RE	Reunion	Department of Reunion	Oversea Department of France	2230.22	861.09
235	Polygon	SE	Seychelles	Republic of Seychelles	UN Member State	222.19	85.79
236	Polygon	KT	Christmas I.	Territory of Christmas Island	Australian Territory	98.77	38.13
237	Polygon	CK	Cocos Is.	Territory of Cocos (Keeling) Islands	Australian Territory	10.07	3.89
238	Polygon	ID	Indonesia	Republic of Indonesia	UN Member State	1847033	713139.44
239	Polygon	TT	Timor-Leste	Democratic Republic of Timor-Leste	UN Member State	15495.72	5982.9
240	Polygon	AS	Australia	Commonwealth of Australia	UN Member State	7694273	2970758.75
241	Polygon	NR	Nauru	Republic of Nauru	UN Member State	19	7.34
242	Polygon	NC	New Caledonia	Territory of New Caledonia and Depende	Non-Self-Governing Territory of France	17946.31	6929.07
243	Polygon	NF	Norfolk I.	Territory of Norfolk Island	Australian Territory	38.97	15.04
244	Polygon	PP	Papua New Guinea	Independent State of Papua New Guinea	UN Member State	458665.63	177090.78
245	Polygon	BP	Solomon Is.	Solomon Islands	UN Member State	21573.21	8329.42
246	Polygon	TV	Tuvalu	Tuvalu	UN Member State	17.49	6.75
247	Polygon	NH	Vanuatu	Republic of Vanuatu	UN Member State	8456.53	3265.07
248	Polygon	MJ	Montenegro	Republic of Montenegro	UN Member State	13773.34	5317.89
249	Polygon	RB	Serbia	Republic of Serbia	UN Member State	87960.13	33961.41

Record: 0   Show: All Selected   Records (0 out of 249 Selected)   Options ▾

(Q12) *The field names in the Identify window are listed in the column starting with OBJECTID. Where are these field names displayed in the table?*

(Q13) *Find the field in the table that represents square miles of land. How many square miles of land does Australia cover?*

(Q14) *Give a brief explanation of the relationship between the Identify window and the table.*

1. Click the Close button that looks like an × at the top right corner of the Identify window. (Be careful not to click the Close button on the ArcMap window by mistake. If you close ArcMap, the map document will close without saving your work.)

Step 8: Explore city data on the world map

1. Turn on the World Cities >100,000 layer.

2. Click the plus sign next to the World Cities >100,000 layer to expand its legend.

□ 🗺 **World Population**
　□ ☑ World Cities > 100,000
　　　Population
　　⚪ 5,000,000 and greater
　　⚪ 1,000,000 to 5,000,000
　　⚪ 500,000 to 1,000,000
　　∘ 250,000 to 500,000
　　· 100,000 to 250,000

Your map displays all the world cities with populations greater than 100,000. There are so many cities that they are all jumbled together on this small map. You need to zoom in to a smaller portion of the world to see distinctions between the cities.

 3. Click the Zoom In tool to select it. (Remember, the button looks pushed in when it is selected.)

The Zoom In tool can be used two different ways. One way to zoom in is to drag a box around the area you want to display. You will zoom in to Europe and Africa.

4. Place your cursor on Greenland. (To find Greenland, move your cursor over the map so the country names display.) Click and hold down the mouse button. Drag the cursor down and to the right. When your cursor is near Australia, release the mouse button.

Another way to zoom in is to click on the place you want to be the center of your map. Now you will zoom in closer to Europe.

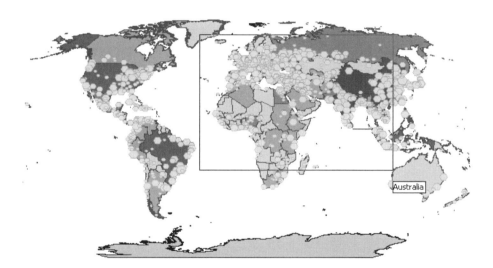

5. Click slowly three times on the yellow dots clustered around Europe. Reposition the magnifying glass to the center of Europe between each click.

> ← Make sure to keep the mouse very still as you click. Otherwise, you may accidentally drag a tiny box and the map will zoom in too much. If this happens, go to the toolbar and click once on the Previous Extent button and do this step again. (You may also use the scroll button on your mouse to zoom in and out of your map.)

You will use the Identify tool to get information about the cities.

6. Click the Identify tool. Click the "Identify from" drop-down menu in the Identify window and select World Cities > 100,000.

7. Click on a red or pink country area that is away from a yellow dot. Notice that the Identify window displays nothing under Value. This is because the Identify tool gets information only for the layer you chose in the list.

8. Click on a yellow dot. Now information appears in the Identify window about the city that you clicked.

(Q15) *Use the Identify tool to find the names and countries of any two cities.*

 9. Click the Close button on the Identify window.

Step 9:   Explore Europe with an attribute table

Earlier in this module you used the Identify tool to learn that a country on the map (Australia) was connected to attribute information about that country. The attributes were displayed in the Identify window.

In GIS, each object on your map is called a feature. For example, Australia is a feature in the World Countries layer, and Paris is a feature in the World Cities >100,000 layer. You will now explore further the connection between features and their attributes.

1. Right-click the World Countries layer in the table of contents and choose Open Attribute Table. If your table is large, drag the right side of the table to the left to make it smaller, like the one pictured below. If your table is covering the map, click its title bar and drag it out of the way until you can see all of Europe.

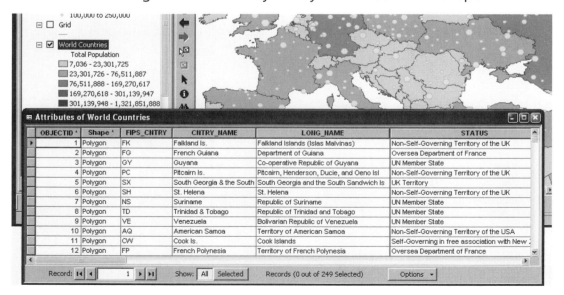

Q16 *What is the name of the table you opened?*

Q17 *What country is listed in the first row of the table?*

Q18 *What country is listed in the last row of the table?*

2. Click the Options button at the bottom of the table and click Find & Replace.

3. Type **Poland** in the "Find what" box in the Find and Replace dialog box. Then click the Find Next button.

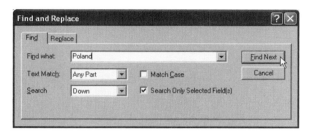

4. Click the Close button to close the Find and Replace dialog box.

5. Click the gray rectangle with the arrow at the beginning of the row for Poland.

	OBJECTID *	Shape *	FIPS_CNTRY	CNTRY_NAME	LONG_NAME
	138	Polygon	AU	Austria	Republic of Austria
	139	Polygon	EZ	Czech Republic	Czech Republic
	140	Polygon	DA	Denmark	Kingdom of Denmark
	141	Polygon	HU	Hungary	Republic of Hungary
▶	142	Polygon	PL	Poland	Republic of Poland
	143	Polygon	LO	Slovakia	Slovak Republic

≡ Attributes of World Countries

Notice that the row in the table and the outline of Poland on the map turn blue to show they are selected.

6. Hold down the Ctrl key on your keyboard. In the table, click the gray boxes for these rows: Slovakia, Belgium, and Germany. (You will need to scroll down several rows to see these countries in the table.)

Q19 *What happens to the map when you click on these rows in the table?*

7. Click the gray box for the United States row. (You may need to scroll down a few rows.)

Q20 *What happens to Poland and the other countries that were highlighted?*

Q21 *Did you see the United States become outlined in blue on the map? If not, why not?*

8. Click the Close button on the table.

9. Click the Full Extent button. The map displays the whole world. You can always click the Full Extent button to get back to a view of the entire map.

(Q22) *Why can you see the United States now but not in the previous step?*

10. Click the Clear Selected Features button.

Whenever you select a feature on your map, it will be outlined in blue to indicate that the feature has been selected. If you want to turn the blue outlines off, you need to clear the selected features.

Step 10: Practice identifying features

1. Click the Zoom In tool.

2. Click and drag a box around the continent of South America. If you would like South America to be larger or smaller, you can use the scroll button on your mouse to zoom in or out.

(Q23) *What do you see on your map?*

3. Click the Identify tool. In the Identify window click the "Identify from" drop-down menu and choose <Top-most layer>.

When you choose Top-most layer, the Identify tool gets information for features in the highest turned-on layer in the table of contents. In this case, if you click a city, you will identify that city; if you click away from a city, you will identify a country.

If the Identify window is covering the map, click its title bar and drag it out of the way.

4. Identify the large, dark red South American country.

(Q24) *What country is it?*

(Q25) *What was this country's total population in 2007?*

5. Identify the large city in the northwest part of this country.

(Q26) *What city is it?*

(Q27) *What population class is this city in?*

6. Look on the map for two cities in Brazil that are in a higher population range. (The yellow symbols for those two cities are larger than the one for the city you just identified.)

🔍 7. Zoom in closer to the two cities to separate them from surrounding cities for identification purposes.

(Q28) *What are the names of these two large cities?*

(Q29) *What population class are these cities in?*

❌ 8. Close the Identify window.

Step 11: Practice zooming out

🔍 1. Click the Zoom Out tool.

2. Drag a two-inch box anywhere on your map.

3. Drag another two-inch box anywhere on your map.

(Q30) *What does your map look like?*

(Q31) *Which button could you use to return your map to full size?*

🌑 4. Zoom your map back to its original full extent.

Step 12: Practice finding a feature

You previously found a country by looking for it in the attribute table. Now you will find a country by looking for it on the map.

1. Turn off the World Cities >100,000 layer.

🔭 2. Click the Find button.

3. Click the white Find box. Type **Sudan**.

4. Click the "In" drop-down menu, scroll down, and click World Countries.

5. Under Search, click the white circle next to "In field."

6. Click the "In field" drop-down menu and select CNTRY_NAME. (You want ArcMap to search for Sudan in the country name attribute field in the World Countries layer.)

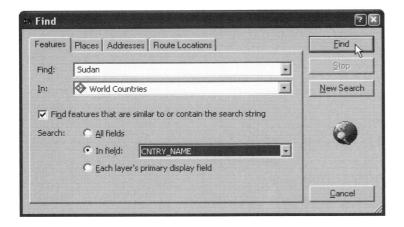

7. Click Find.

   A results list appears at the bottom of the Find window, and Sudan is listed. At the bottom of the window there is a message telling you "One object found."

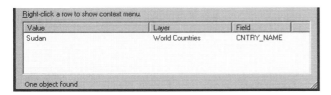

   If you get a results list that is blank and the message at the bottom of the window says "No objects found," go back to steps 3–7. Make sure you spell the country name correctly and enter the correct information.

8. Move the Find window off the map.

9. Right-click on Sudan in the results list and click Flash. The country of Sudan flashes dark green in the map. (If you didn't see it flash, repeat this step and watch Africa on the map.)

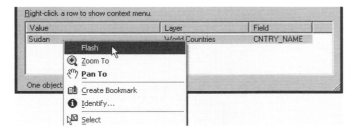

10. Right-click Sudan in the results list again and click Identify.

   All or part of the Identify window may be hidden behind the Find dialog box. If so, move the Find dialog box out of the way.

(Q32) *How many square kilometers in area is Sudan?*

(Q33) *How many people lived in Sudan in 2007?*

(Q34) *Compute the number of people per square kilometer. (Divide population by square kilometers and round off to two decimal places).*

(Q35) *Does the number of people per square kilometer seem low or high?*

11. Close the Identify window.

Step 13: Zoom to a feature and create a bookmark

1. Click the New Search button in the Find window.

2. Type **Qatar** in the Find box.

3. Click Find.

4. Right-click Qatar in the results list. Click Select.

   Qatar is outlined in blue on the map.

(Q36) *Is Qatar a large country or a small one?*

5. Right-click Qatar in the results list and choose Zoom To. This zooms in and centers the map on Qatar.

   You will now create a bookmark for Qatar that you will later use to quickly return to this map extent.

6. Right-click Qatar in the results list and choose Create Bookmark.

7. Click the View menu and point to Bookmarks. Notice that the bookmark you created for Qatar is listed. You will come back to this later.

8. Right-click Qatar in the results list in the Find box and click Identify.

(Q37) *How many people lived in Qatar in 2007?*

(Q38) *How many cell phones did they have? (Scroll down to MOBPHNS.)*

You may notice that the cell phone data is for the year 2006 and not 2007 as the population data. Many times data is collected at different times, and you will notice this type of disparity (difference) in the data. You do not need to worry about this disparity when using these two datasets to answer the following question.

Q39 *How many people were there for every cell phone in Qatar? (Divide the population of Qatar by the number of cell phones in Qatar. Don't worry about the disparity between the population year and the year that phone line data was collected.)*

9.   Close the Identify and Find windows.

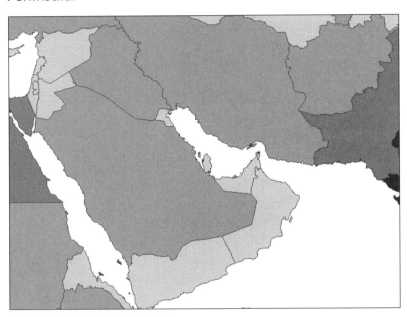

10. Click three times on the Fixed Zoom Out button. (Its arrows point outward.)

Q40 *What large country is directly west of Qatar? (Use MapTips)*

11. Click the Fixed Zoom Out button nine more times until you see the entire Arabian Peninsula.

Step 14: Explore the World Population map further

Whenever you wish to center the map differently, you can use the Pan tool to move the map around.

1.   Click the Pan tool.

2.   Use MapTips to locate the country of Egypt at the left edge of the map.

3.   Click and hold Egypt. Drag the hand diagonally to the bottom right corner of the map area. Release the mouse button.

Q41 *What boot-shaped country do you see on the map?*

Q42 *What was the 2007 population of that country?*

Q43 *How many cell phones did that country have?*

Q44 *How many people were there for every cell phone in that country? (Divide the population by the number of cell phones for the country.)*

Q45 *How many cell phones were there per person in that country?*

4. Pan east to Japan by using the Pan tool (hand).

Q46 *What was the 2007 population of Japan?*

Q47 *How many cell phones did Japan have?*

Q48 *How many people were there for every cell phone in Japan? (Divide the population of Japan by the number of cell phones in Japan.)*

5. Close the Identify window.

6. Click the View menu. Point to Bookmarks and click Qatar. The map returns to Qatar using the bookmark you saved.

7. Click the Clear Selected Features button.

Q49 *What happened to Qatar?*

Step 15: Use the What's This? tool

If you forget what a particular button is for or how to use a tool, you can ask ArcMap to help you.

1. Click the What's This? tool on the Standard toolbar. (Notice your cursor turns into an arrow with a question mark.)

2. Click the Find button. An explanation of the Find button appears.

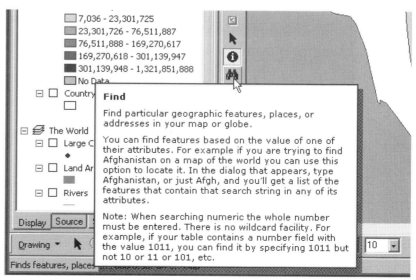

3. Click anywhere inside the pop-up message box to make it disappear.

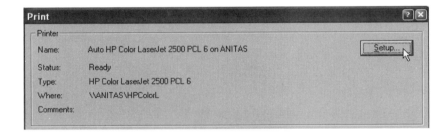4. Click the What's This? tool.

5. Click any other button you want to know about.

6. When you are finished looking at the help, click the pop-up message box to close it.

Step 16: Label and print a map

You will now learn an easy way to label the countries and to print the map.

1. Use the Pan and Zoom tools to focus the map on a location of your choice.

2. Right-click World Countries and choose Label Features. Country names are added to the map.

(Q50) *Where do you think these labels come from?*

3. Click the File menu and then click Print.

4. In the Print dialog box, click the Setup button.

Print	
Printer	
Name:	Auto HP Color LaserJet 2500 PCL 6 on ANITAS	Setup...
Status:	Ready
Type:	HP Color LaserJet 2500 PCL 6
Where:	\\ANITAS\HPColorL
Comments:	

5. Under Printer Setup in the Page and Print Setup dialog box, select the printer specified by your teacher.

6. Under Map Page Size, check the box next to Use Printer Paper Settings.

7. Under Paper (the section above Map Page Size), make sure the paper size is set to Letter. (You may need to select it from the drop-down menu.)

8. Click Portrait if you want the top of the map to align with the short edge of the paper. Click Landscape if you want the top to align with the long edge.

9. At the bottom of the dialog box, check the box next to Scale Map Elements proportionally to changes in Page Size. Your dialog box should look similar to the one pictured on the following page.

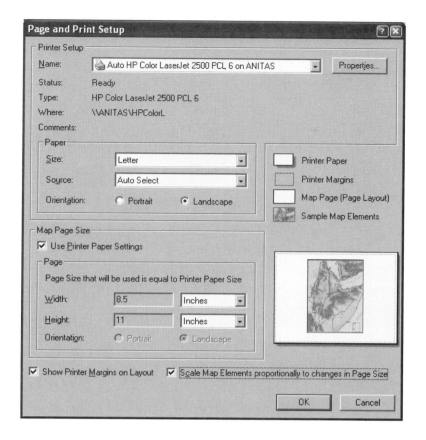

10. Click OK on both the Page and Print Setup window and the Print window.

Step 17: Close the map document and exit ArcMap

1. Click the File menu.

2. Click Exit.

3. Click No in the box that asks whether you want to save changes.

Answer sheet
Module 1, Lesson 1

The basics of ArcMap

Step 3: Work with layers

Q1)   Which layers are not visible on the map but are turned on in the table of contents?

_____

_____

Q2)   What happened on your map?

_____

Q3)   What happened on your map?

_____

Q4)   What would happen if you dragged Rivers under Land Areas?

_____

_____

Step 4: Change the active data frame

Q5)   What is the name of the layer that is turned on in the World Population data frame?

_____

Step 6: Obtain information about a country

Q6)   What is the fourth listing under Field? _____

Q7)   What is the fifth listing in this column? _____

Q8)   What is the final listing in this column? (You will need to scroll down.)

_____

Q9) What do you guess the field SQMI stands for?

_____

Q10) What is the number to the right of the field SQMI?

_____

Step 7: Compare Identify data with table data

Q11) Which row in the table has the attributes for Australia? _____

Q12) Where are these field names displayed in the table?

_____

Q13) How many square miles of land does Australia cover?

_____

Q14) Give a brief explanation of the relationship between the Identify window and the table.

_____

_____

_____

Step 8: Explore city data on the world map

Q15) Use the Identify tool to find the names and countries of any two cities.

City	Country

Step 9: Explore Europe with an attribute table

Q16) What is the name of the table you opened?

_____

Q17) What country is listed in the first row of the table?

_____

Q18) What country is listed in the last row of the table?

_____

Q19) What happens to the map when you click on these rows in the table?

_____

Q20) What happens to Poland and the other countries that were highlighted?

_____

_____

Q21) Did you see the United States become outlined in blue on the map? If not, why not?

_____

_____

_____

Q22) Why can you see the United States now but not in the previous step?

_____

_____

_____

Step 10: Practice identifying features

Q23) What do you see on your map?

_____

Q24) What country is it? _____

Q25) What was this country's population in 2007? _____

Q26) What city is it? _____

Q27) What population class is this city in? _____

Q28) What are the names of these two large cities?

_____

Q29) What population class are these cities in? _____

Step 11: Practice zooming out

Q30) What does your map look like?

_____

Q31) Which button could you use to return your map to full size?

_____

Step 12: Practice finding a feature

Q32) How many square kilometers in area is Sudan? _____

Q33) How many people lived in Sudan in 2007? _____

Q34) Compute the number of people per square kilometer. _____

Q35) Does the number of people per square kilometer seem low or high? _____

Step 13: Zoom to a feature and create a bookmark

Q36) Is Qatar a large country or a small one? _____

Q37) How many people lived in Qatar in 2007? _____

Q38) How many cell phones did they have? _____

Q39) How many people were there for every cell phone in Qatar? _____

Q40) What large country is directly west of Qatar? _____

Step 14: Explore the World Population map further

Q41) What boot-shaped country do you see on the map? _____

Q42) What was the 2007 population of that country? _____

Q43) How many cell phones did that country have? _____

Q44)   How many people were there for every cell phone in that country? _____

Q45)   How many cell phones were there per person? _____

Q46)   What was the 2007 population of Japan? _____

Q47)   How many cell phones did Japan have? _____

Q48)   How many people were there for every cell phone in Japan? _____

Q49)   What happened to Qatar? _____

Step 16: Label and print a map

Q50)   Where do you think these labels come from?

_____

_____

Module 1, Lesson 2

The geographic inquiry process

- Supplement: The geographic inquiry process
- Activity
- Answer sheet
- Assessment

Supplement

The geographic inquiry process

Step	What to do	Examples from module 1, lesson 2
Ask a geographic question	Think about a topic or place and identify something interesting or significant about it. Turn that observation into a geographic question or hypothesis that you can investigate. Types of geographic questions include: • Where are things located? • How do things change from one place to another? • Why do things change from one place to another?	• What countries have the most and the fewest phone lines? • Does the number of phone lines vary proportionately with the number of people among the world's most populous countries? • Why does Pakistan have more people per phone line than India?
Acquire geographic resources	Identify the information and data needed to answer your geographic question: • What is the geographic focus of your research? • For which time frame do you need data? • On what specific topics do you need data?	Use data and maps from the module 1 data folder that contains: • World countries • Population data for world countries • Phone lines data for world countries
Explore geographic data	Turn the data into maps, tables, and graphs and look for patterns in the way things change from one place to another. Some ways to explore data in ArcMap include the following: • Create a map document and add data layers. • Turn layers on and off. Zoom and pan the map. Look at individual features and what surrounds them. • Change the symbols used to represent features. • Look for ways features in one layer relate to features in other layers.	Explore map layers and attribute tables containing population and phone line data for world countries.
Analyze geographic information	Focus on the information and maps that most seem to answer your questions. For example: • Find or identify particular features. • Select features with specific attributes to meet specific criteria. • Calculate new attributes from existing ones to get new information. • Draw conclusions from what you have seen in the maps, tables, and graphs and answer your geographic question.	• Calculate phone line densities for all countries. • Research and record population, phone line, and phone line density information for selected countries. • Rank selected countries by population and phone line density and compare the two lists. • Compare your answer with your initial hypothesis.
Act on geographic knowledge	Your conclusions are the result of turning pieces of data into geographic knowledge. Think about how you could share this knowledge or how you or someone else could use it to make a decision, correct a problem, or help others.	Devise a plan of action for the phone system in one of the countries that you researched.

Module 1, Lesson 2

The geographic inquiry process

Much like scientific analysis, geographic inquiry involves a process of asking questions and looking for answers. The geographic inquiry process consists of the following five steps:

In this activity you will learn how maps and GIS can help you in the geographic inquiry process. At the same time, you will be practicing the ArcGIS skills you learned in Lesson 1 and learning some new ones.

Step 1: Open a map document

1. Double-click the ArcMap icon on your computer's desktop.

2. When the ArcMap start-up dialog box appears, click **An existing map** and click OK.

3. Navigate to the module 1 folder (**OurWorld2\Mod1**) and choose **Module1.mxd** (or **Module1**) from the list.

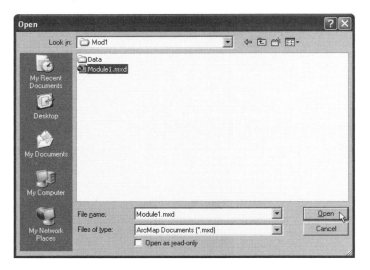

4. Click Open.

Step 2: Activate the World Population data frame

1. Enlarge the ArcMap window by stretching it.

2. In the table of contents, click the minus sign next to The World data frame to collapse it.

3. Click the plus sign next to World Population. The table of contents expands to show the layers in the World Population data frame.

4. Right-click World Population and left-click Activate. The World Population data frame becomes bold and a new map displays in the map area. (You may need to widen the table of contents so that the World Countries legend is not cut off.)

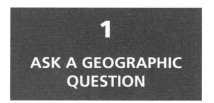

1

ASK A GEOGRAPHIC QUESTION

Step 3: Ask a geographic question and develop a hypothesis

You might look at a map and then think of a question that it might help you answer. Or you might think of the question first and then look for maps or GIS layers that might help you answer the question.

In this lesson you will try to answer the following geographic question: Do the number of phone lines vary proportionately with the number of people among the world's most populous countries?

Answers to questions in this activity should be recorded on the answer sheet.

Q1 *What makes this a geographic question?*

Q2 *Write a hypothesis that answers the geographic question, as you would for any scientific inquiry.*

A hypothesis is an educated guess. Your hypothesis may be right or wrong in the end. The goal is not to know the answer before you start your research but to have a kind of "ballpark idea" or hunch about what the answer might be. Your research will help you determine whether your hypothesis is likely true or false.

2
ACQUIRE GEOGRAPHIC RESOURCES

Step 4:   Add a layer to your map

Next, you need to identify the kind of information that will help you answer your question. You want to display this data as layers in your ArcGIS map.

Q3 *Your map already has a layer with world countries and their populations. What other attribute of countries do you need in order to investigate your hypothesis?*

1. Click the Add Data button.

2. Click the Connect To Folder button. Navigate to the OurWorld2 folder (**OurWorld2**). Click OK. The connection is added to the list of locations in the Add Data dialog box. (If a folder connection to the OurWorld2 folder already exists on your computer, you may skip this step.)

3. Navigate to the Mod1 LayerFiles folder (**OurWorld2\Mod1\Data\ LayerFiles**). Click **World Phone Lines Land.lyr**.

4. Click Add.

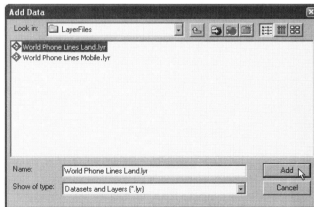

Q4 *What is the name of the layer that has been added to your table of contents?*

An ArcMap layer file contains the complete definition of a layer, including its name, data source, symbology, and other properties. You can save a layer outside a map document as a layer file so it can be reused in other maps.

> ## 3
> ### EXPLORE GEOGRAPHIC DATA

Step 5: Explore the World Phone Lines Land map

1. Turn off the World Countries layer.

☐ ☑ World Phone Lines Land
      PHONLNS
      ☐ 1 - 2,572,000
      ☐ 2,572,001 - 10,396,100
      ☐ 10,396,101 - 25,049,000
      ☐ 25,049,001 - 58,780,000
      ☐ 58,780,001 - 350,433,000
      ☐ No Data

Q5 *What color in the legend for the World Phone Lines Land layer indicates countries with the fewest phone lines?*

Q6 *What color indicates countries with the most phone lines?*

Q7 *What color indicates countries with no data available for this layer?*

Notice that the colors in the legend change gradually from lightest to darkest. The name for this type of legend is a graduated color legend.

Q8 *What other layer in your map has a graduated color legend?*

To answer the following questions, you may need to turn layers on or off and you may need to use MapTips or the Identify tool to find out a country's name.

Q9 *Which two countries have the most phone lines?*

Q10 *On which continent are most of the countries with the fewest phone lines?*

(Q11) *Which two countries have the largest populations?*

(Q12) *Name the two countries that are in the same population class (color) as the United States.*

(Q13) *Which of the two countries, if any, are in the same phone line class (color) as the United States?*

2. With the Identify tool, click on the United States.

3. Read the geographic question again.

   Geographic question: Do the number of phone lines vary proportionately with the number of people among the world's most populous countries?

4. Scroll down slowly in the Identify window and look at the different fields.

4

ANALYZE GEOGRAPHIC INFORMATION

(Q14) *What two fields might help in answering the geographic question?*

5. Close the Identify window.

Step 6: Investigate the relationship between phone lines and population in China

1. Click the Full Extent button on the Tools toolbar. (This will return the map to its original size if you have zoomed or panned the map.)

2. To unselect (turn the blue outline off) any countries that might be selected, click the Clear Selected Features button. (If the button is disabled, or grayed out, it means that there are no features currently selected in the World Population data frame.)

3. Click the Find button and type China in the Find dialog box.

4. Click Find. China appears in the results list at the bottom of the Find window.

5. Right-click on China in the results list and click Identify. (If you get more than one result for China, right-click on the row for World Phone Lines Land.)

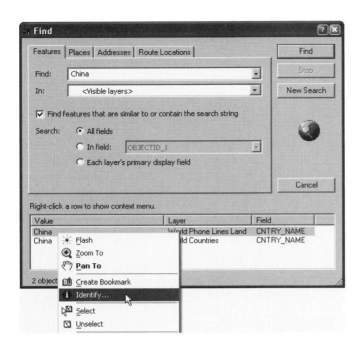

Q15 *What was the population of China in 2007? (You may need to move the Find dialog box out of the way.)*

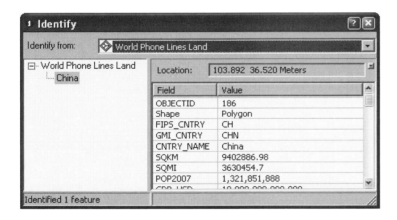

Q16 *How many phone lines did China have in 2007? (Scroll down and look for the PHONLNS attribute.)*

Remember not to worry about the disparity between the population year and the year that phone line data was collected.

6. Close the Find and Identify windows.

7. Click Start, Programs, Accessories, Calculator.

8. Divide the population of China by the number of phone lines.

Q17 *What was the number of people per phone line in China? (Round off the result to the nearest two decimal places.)*

Step 7: Calculate the numbers of people per phone line for all countries

1. In the table of contents, right-click World Phone Lines Land and click Open Attribute Table.

2. Scroll all the way to the right until you see the last field in the table, PHON_DEN (for phone density). Notice that the values in this field are Null, indicating that no value has been assigned.

3. Right-click on the PHON_DEN field heading and click Field Calculator.

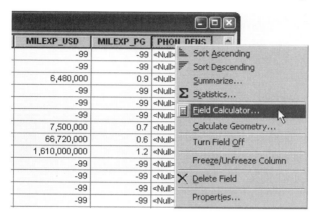

4. In the Field Calculator message box, click Yes to continue.

    You want to divide the population values in the POP2007 field by the number of phone lines in the PHONLNS field.

5. In the Fields list, double-click POP2007. It now appears in the white text box below.

6. Click the division button.

7. In the Fields list, find PHONLNS and double-click it.

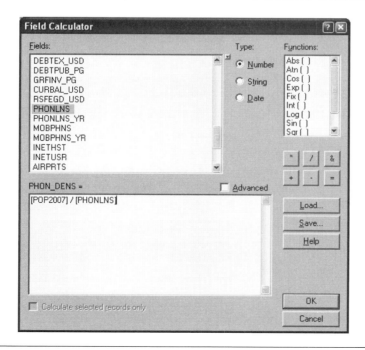

If you make a mistake, highlight and delete any text in the white box.

8. Click OK.

   The field calculator calculates the number of people per phone line and places the values in the PHON_DEN field in the attribute table.

9. Scroll down in the table to see all the PHON_DEN values.

   You will see some phone density values of 1 or a negative number. These phone density numbers are not meaningful because they were computed using a null or no data value (-99) in the POP2007 or PHONLNS field. For this exercise, you can ignore those numbers.

10. Close the attribute table.

Step 8: Investigate the relationship between population and phone lines for all countries

1. Click the Find button. China appears in the Find dialog box.

2. Click Find.

3. Right-click on China in the results list and click Identify.

   (Q18) *What is the number of people per phone line (PHON_DEN) for China?*

   (Q19) *Does this number agree with the value you calculated in Q17?*

   (Q20) *For each country in the table on the answer sheet, what is the population, number of phone lines, and number of people per phone line? (Use the Find and Identify tools.)*

4. Click the File menu, then Exit, and then No in the box that asks whether you want to save changes.

Step 9: Analyze the results of your research

   (Q21) *In the table on the answer sheet, the column on the left ranks the countries by population from highest to lowest. In the column on the right, rank the countries from the lowest number of people per phone line to the highest number of people per phone line, using the data from Q20. Then draw lines connecting the name of each country in one column with the name of the same country in the other column.*

(Q22) *Which country has the fewest people per phone line? What is the number of people per phone line in this country?*

(Q23) *How does the country in Q22 rank in population size among the eight countries in your table?*

(Q24) *Which country has the most people per phone line? What is the number of people per phone line in this country?*

(Q25) *How does the country in Q24 rank in population size among the eight countries in your table?*

(Q26) *What is the population of Japan? What is the number of people per phone line in Japan?*

(Q27) *What country has the most phone lines? How does the number of people per phone line in this country compare with the numbers of people per phone line in the seven other countries in your table?*

(Q28) *Russia and Pakistan have about the same number of people. Why do you suppose these two countries have such different numbers of people per phone line? What factors do you think contribute to this disparity?*

(Q29) *What is the answer to the geographic question: Do the number of phone lines vary proportionately with the number of people among the world's most populous countries?*

(Q30) *Compare your initial hypothesis (Q2) with your answer in Q29. How does your hypothesis compare with your answer to the geographic question?*

5
ACT UPON GEOGRAPHIC KNOWLEDGE

Step 10: Develop a plan of action

The last step of the geographic inquiry process is to act on what you have learned. Your action plan might be simply to repeat the process; thinking about what you've learned often leads to deeper, more interesting geographic questions.

Imagine that you are an expert specializing in telecommunications. You need to devise a plan for China, Brazil, Indonesia, or the United States that deals effectively with the basic concern of your original geographic question.

To develop an effective plan, you may need to conduct further research on the phone system within your country. If you decided, for instance, that increasing the number of phone lines operating in your chosen country would improve the quality of life there, you could come up with a written plan of action for telecommunications officials, pointing out strengths and weaknesses, and explaining where and why expansion would be most beneficial.

Q31 *What country did you choose? Use the information in the table in Q20 to describe the phone line situation in your chosen country.*

Q32 *Do you think that increasing the number of phone lines operating in your chosen country would improve the quality of life there? Why or why not?*

Q33 *List three concerns you have about increasing the number of phone lines in your chosen country.*

Q34 *List two new geographic questions that you would like to investigate to help you develop a sound plan.*

Name_____   Date_____

Module 1, Lesson 2

The geographic inquiry process

Step 3: Ask a geographic question and develop a hypothesis

Q1)   What makes this a geographic question?

_____

_____

Q2)   Write a hypothesis that answers the geographic question.

_____

_____

_____

_____

_____

Step 4: Add a layer to your map

Q3)   What other attribute of countries do you need in order to investigate your hypothesis?

_____

Q4)   What is the name of the layer that has been added to your table of contents?

_____

Step 5: Explore the World Phone Lines Land map

Q5)   What color in the legend for the World Phone Lines Land layer indicates countries with the fewest phone lines? _____

Q6) What color indicates countries with the most phone lines?

_____

Q7) What color indicates countries with no data available for this layer?

_____

Q8) What other layer in your map has a graduated color legend?

_____

Q9) Which two countries have the most phone lines?

_____

Q10) On which continent are most of the countries with the fewest phone lines?

_____

Q11) Which two countries have the largest populations?

_____

Q12) Name the two countries that are in the same population class (color) as the United States.

_____

Q13) Which of the two countries, if any, are in the same phone line class (color) as the United States?

_____

Q14) What two fields might help in answering the geographic question?

_____

Step 6: Investigate the relationship between phone lines and population in China

Q15) What was the population of China in 2007? Record the answer in the table below.

Q16) How many phone lines did China have in 2007? Record the answer in the table below.

Q17) What was the number of people per phone line in China? Record the answer in the table below.

Country	Population	Phone lines	People/phone line
China			

Step 8: Investigate the relationship between population and phone lines for all countries

Q18) What is the number of people per phone line (PHONE_DEN) for China?

_____

Q19) Does this number agree with the value you calculated in Q17?

_____

Q20) For each country in the table below, what is the population, number of phone lines, and number of people per phone line? The first country, China, is already filled in for you.

Country	Population	Phone lines	People/phone line
China	1,321,851,888	350,433,000	3.77
India			
United States			
Indonesia			
Brazil			
Pakistan			
Russia			
Japan			

Step 9: Analyze the results of your research

Q21) In the table below, the column on the left ranks the countries by population from highest to lowest. In the column on the right, rank the countries from the lowest number of people per phone line to the highest number of people per phone line, using the data from Q20. Then draw lines connecting the name of each country in one column with the name of the same country in the other column.

Ranked by population (highest to lowest)	Ranked by people/phone line (lowest to highest)
China	
India	
United States	
Indonesia	
Brazil	
Pakistan	
Russia	
Japan	

Q22) Which country has the fewest people per phone line? _____

What is the number of people per phone line in this country? _____

Q23) How does the country in Q22 rank in population size among the eight countries in your table? _____

Q24) Which country has the most people per phone line? _____

What is the number of people per phone line in this country? _____

Q25) How does the country in Q24 rank in population size among the eight countries in your table? _____

Q26) What is the population of Japan? _____

What is the number of people per phone line in Japan? _____

Q27) What country has the most phone lines? _____

How does the number of people per phone line in this country compare with the numbers of people per phone line in the seven other countries in your table?

_____

Q28) Russia and Pakistan have about the same number of people. Why do you suppose these two countries have such different numbers of people per phone line?

_____

_____

What factors do you think contribute to this disparity? _____

_____

_____

_____

Q29) What is the answer to the geographic question?

_____

_____

_____

_____

Q30) Compare your initial hypothesis (Q2) with your answer in Q29. How does your hypothesis compare with your answer to the geographic question?

_____

_____

_____

_____

Step 10: Develop a plan of action

Q31) What country did you choose? _____

Use the information in the table in Q20 to describe the phone line situation in your chosen country.

_____

_____

_____

_____

Q32) Do you think that increasing the number of phone lines operating in your chosen country would improve the quality of life there? Why or why not?

_____

_____

_____

_____

Q33) List three concerns you have about increasing the number of phone lines in your chosen country.

_____

_____

_____

Q34) List two new geographic questions that you would like to investigate to help you develop a sound plan.

_____

_____

_____

Middle school assessment
Module 1, Lesson 2

The geographic inquiry process

Open the ArcMap document Module1.mxd (or Module1). Use the ArcMap skills you have learned in this lesson to do the following things:

1. Create a map with at least three different layers.

2. Zoom in on the map to an area of your choice.

3. Find out three pieces of specific information about the area you chose (use the Identify button).

4. Write a geographic question that involves one of the pieces of information that you listed in item 3.

5. Compare your area, or one country within your area, to either China or India. List characteristics that the two places have in common and characteristics that are different.

6. Print the map and attach it to this page.

High school assessment
Module 1, Lesson 2

The geographic inquiry process

Open the ArcMap document Module1.mxd (or Module1). Use the ArcMap skills you have learned in this lesson to do the following things:

1. Create a map with at least three different layers.

2. Zoom in on the map to an area of your choice.

3. Find out three pieces of specific information about the area you chose (use the Identify button).

4. Write a geographic question that involves one of the pieces of information that you listed in item 3.

5. Write a brief paragraph explaining what you learned about the geography of the area. How might these characteristics be reflected in the social, cultural, or economic character of this place?

6. Compare your area, or one country within your area, to either China or India. Describe characteristics that the two places have in common and characteristics that are different.

7. Print the map and attach it to this page.

Module 2, Lesson 1

The earth moves

A global perspective

- Supplement: Map exercise
- Supplement map
- Activity
- Answer sheet
- Assessment
- Assessment map

Supplement

Map exercise

On the map on the next page, mark with the letter *V* eight potential volcano sites and mark with the letter *E* eight locations where you think earthquakes typically occur.

Answer the following questions as a group:

1.   *What similarities do you see between the maps of the group members?*

2.   *How did the members of your group choose the locations they marked?*

3. *At what locations were earthquake and volcano sites marked close together by the members of your group?*

_____

_____

_____

_____

_____

_____

_____

4. *What major cities are close to these locations?*

_____

_____

_____

_____

_____

_____

_____

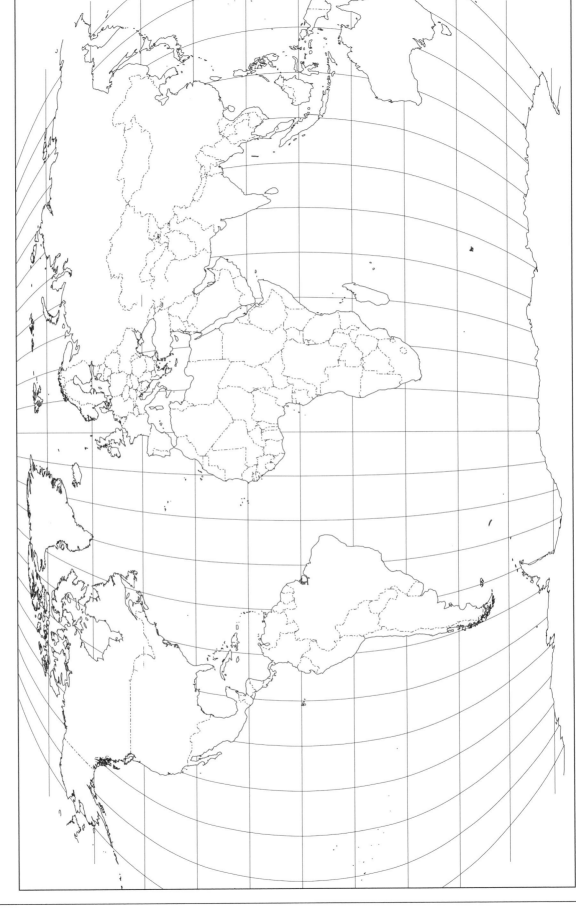

The World

Supplement map

M2
L1

Module 2, Lesson 1

The earth moves

In this activity, you will observe worldwide patterns of seismic activity (earthquakes) and volcanic activity (volcanoes). You will analyze the relationships of those patterns to tectonic plate boundaries and major physical features of the earth's surface. Then you will identify cities at risk.

Step 1:   Open a map document

1.   Double-click the ArcMap icon on your computer's desktop.

2.   When the ArcMap start-up dialog box appears, click **An existing map** and click OK.

3. Navigate to the module 2 folder (**OurWorld2\Mod2**) and choose **Global2.mxd** (or **Global2**) from the list.

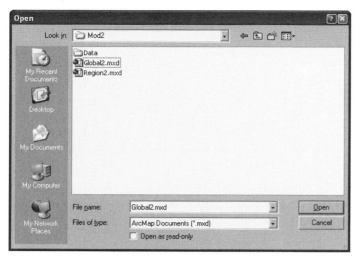

4. Click Open.

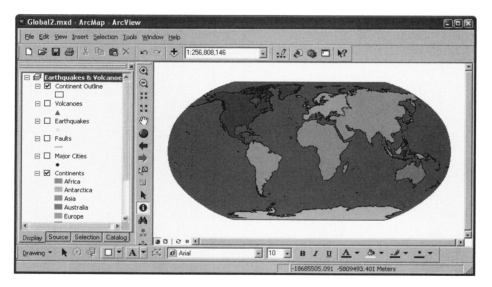

When the map document opens, you see a map with three layers turned on: Continent Outline, Continents, and Ocean. The check mark next to the layer name tells you the layer is turned on and visible on the map.

Step 2:   Analyze earthquake locations

You will now compare the predictions you made in the Supplement about earthquake and volcano locations to actual data using GIS.

1. Turn on the Earthquakes layer by clicking the box to the left of the name in the table of contents.

This places a check mark in the box, and earthquake points are drawn on the map. The points show the locations of earthquakes that occurred between January 2004 and April 2007.

Q1 *Do earthquakes occur in the places you predicted? List the regions you predicted correctly for earthquake locations.*

Q2 *What patterns do you see on the map?*

Answers to questions in this activity should be recorded on the answer sheet.

Step 3: Sort and analyze earthquake magnitudes

You can take a closer look at the data behind the dots by looking at the attribute table of the Earthquakes layer. An attribute table contains specific information about the features in a layer. In the Earthquakes layer, each point represents an earthquake with a magnitude of 3.5 or greater on the Richter scale. You will focus on the 20 strongest earthquakes.

1. In the table of contents, right-click Earthquakes and click Open Attribute table. You see all the attribute data associated with the yellow earthquake points on the map. (Do not maximize this table—it will prevent you from viewing the map at the same time.)

2. Scroll down to see more records. Remember, each record in this table represents one point on the map.

3. Click the field (column heading) labeled MAGNITUDE to select it. This field represents the magnitude of the earthquakes.

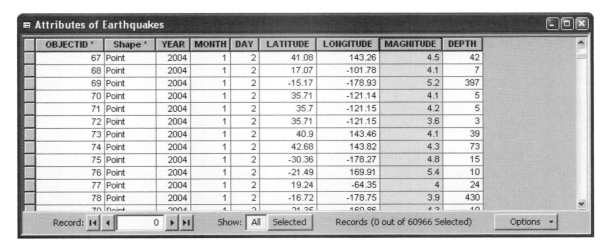

Now you will put the magnitudes in order from largest to smallest.

4. Scroll up to the top of the table. Right-click the MAGNITUDE field heading and click Sort Descending. The records are rearranged from largest to smallest.

Now you will select the 20 largest earthquakes.

5. Hold down the Ctrl key, click the small gray box to the left of the first record in the table, and drag your mouse until the first 20 records are highlighted in blue.

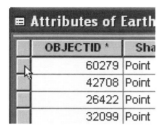

To make sure you have highlighted 20 earthquakes, look at the status bar at the bottom of the table. It should look like the following graphic:

Records (20 out of 60966 Selected)

> If you select too many records, click the Options button at the bottom of the table, click Clear Selection, and try again. You can stretch the bottom of your table to make it easier to select 20 earthquakes.

When you select a record in the attribute table, its point on the map will be highlighted also.

6. Move the attribute table out of the way so you can see where the 20 strongest earthquakes are located on the map.

Q3 *How do the 20 selected locations compare to your Supplement map? List three ways.*

7. At the bottom of the attribute table, click the Options button and click Clear Selection.

> If you don't see the Options button, your table window may be too narrow. Widen your table window until you can see the Options button near the bottom right corner of the table.

8. Close the attribute table.

Step 4: Analyze volcano data

1. Turn off the Earthquakes layer and turn on the Volcanoes layer.

```
☐ ⬚ Earthquakes & Volcanoes
    ☐ ☑ Continent Outline
        ☐
    ☐ ☑ Volcanoes
        ▲
    ☐ ☐ Earthquakes
    ☐ ☐ Faults
        ⌐
```

Q4 *How do the volcano locations compare with your original predictions? List the regions of volcanic activity you predicted correctly.*

Q5 *What patterns do you see in the volcano locations, and how do they compare with the earthquake patterns? (Turn the Earthquakes layer on and off as needed.)*

The data includes volcanoes that are not active. You will focus on the active volcanoes.

Step 5:   Select all active volcanoes

1. In the table of contents, right-click Volcanoes and click Open Attribute Table.

   The Type field tells you if a volcano is active, potentially active, or solfatara (emits gases but is otherwise inactive).

2. Right-click the Type field heading and click Sort Ascending. Scroll down and you will notice that there are many active volcanoes.

   It would be difficult to highlight all of these, as you did with the Earthquakes layer. Instead, you will select all of the active volcano records at once.

3. At the bottom of the attribute table, click the Options button and click Select By Attributes.

4. You will complete the Select by Attributes dialog box working from top to bottom.
   * Make sure Method is set to "Create a new selection."
   * Double-click Type in the list of fields.
   * Click the equals sign (=).
   * On the right-hand side, click the Get Unique Values button. Three values appear in the list of unique values. Double-click "Active" in this list.

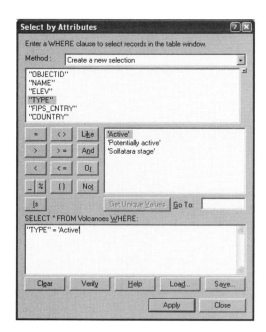

The information now appearing in the lower area of the dialog box ("TYPE" = 'Active') is known as a query expression.

5. Click the Verify button near the bottom of the dialog box. If the expression is successfully verified, click OK.

> If you receive a syntax error, check that your equation is exactly like the one in the picture above. If it isn't, click the Clear button and try again.

6. At the bottom of the dialog box, click Apply. All the active volcanoes are selected and highlighted blue, both in the table and on the map.

7. Close the Select by Attributes dialog box. Close the attribute table to see the map. Use the Zoom and Pan tools or the middle scroll button on your mouse to explore where the active volcanoes are located.

Q6 *What pattern do you see?*

Q7 *Formulate a hypothesis as to why volcano eruptions and earthquakes happen where they do.*

Step 6: Identify active volcanoes on different continents

1. Click the Identify tool. The Identify window displays. Click in the "Identify from" drop-down menu and select Volcanoes from the list.

2. Move your cursor over the map display. Notice how the cursor has an "i" next to it.

3. Click an active volcano on the map. The Identify window shows you the name of the volcano, its elevation, type, and country. For example:

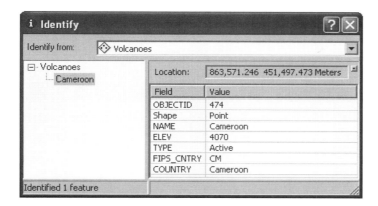

4. Move the Identify window so you can see your map. Using the scroll button on your mouse, zoom in to the continent of your choice.

Remember, you can zoom in and out by scrolling your mouse scroll button, and you can pan your map by holding down the scroll button and moving your mouse.

Q8 *Use the Identify tool to find the names, elevations, and countries of three active volcanoes.*

5. Close the Identify window.

6. Click the Clear Selected Features button to unselect all of the active volcanoes.

7. Click the Full Extent button to see the entire world on the map.

Step 7: Add the plate boundaries layer

The earth is always changing. The crust of the earth is composed of several tectonic plates that are always on the move. The effects of movement are most noticeable at the boundaries between the plates.

Q9 *Based on the locations of earthquakes and volcanoes, where do you think the plate boundaries are? Draw them on the Supplement map.*

You will now investigate the locations of plate boundaries and their effects on adjacent physical features. The four basic types of plate boundaries are:

- **Divergent boundary.** One or two plates are splitting apart. New crust is being formed from the center of the earth, causing the plate to spread. Rift valleys are one example of this type of plate movement.

- **Convergent boundary.** Two plates are colliding, forcing one plate to dip down underneath another one. The plate that is folding under has old crust that is being destroyed, while the plate on top has mountains and volcanoes being formed. In the ocean, these appear as trenches.
- **Transform boundary.** Plates are sliding against each other, causing large fault lines and mountains to form. Here, the crust is neither created nor destroyed.
- **Plate boundary zones (zigzagged).** Plate boundaries appear erratic (zigzagged). Scientists believe there are microplates in these areas, but it is unclear what effect they have on the physical environment.

1. Turn off the Volcanoes layer.

2. Click the Add Data button.

3. Navigate to the module 2 Data folder (**OurWorld2\Mod2\Data**). Double-click **World2.gdb** to open it. Click **plates**.

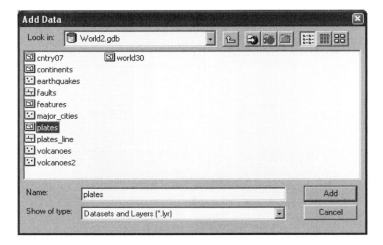

4. Click Add. The plates layer is added to your table of contents.

5. In the table of contents, click the symbol beneath plates. The Symbol Selector opens.

6. On the right side, click the Fill Color drop-down arrow and click No Color. This will create an outline of the plate boundaries.

7. Click the Outline Color drop-down arrow. Pause your cursor over a color to see its name. Click the Electron Gold color.

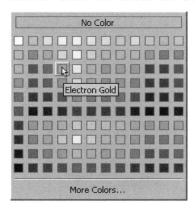

8. Increase the Outline Width to 2.

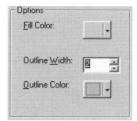

9. Click OK. The gold outline symbol appears in the table of contents and on the map.

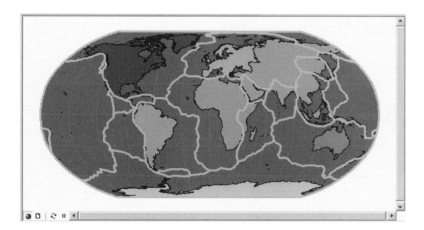

(Q10) *Compare the actual plate boundaries to the ones you drew on the Supplement map. Record all similarities and differences.*

Step 8: Add a layer file and an image

In order to get a closer look at physical features and plate boundaries, you will add two more layers:
- **Major Physical Features.lyr:** a layer showing major landforms and ocean floor features of the planet
- **Earth_wsi.sid:** a color-shaded relief map of the earth made from a satellite image

 1. Click the Add Data button and then click the Up One Level button to navigate to the Data folder. Double-click the LayerFiles folder to open it.

2. Double-click **Major Physical Features.lyr** to add it to your map.

 3. Click the Add Data button and then click the Up One Level button to navigate to the Data folder. Double-click the Images folder to open it.

4. Click **earth_wsi.sid** once and click Add (a file with a .sid extension is a compressed image file).

5. In the table of contents, click earth_wsi.sid and drag it above the Continents layer.

(Q11) *Are there any areas where physical features, plate boundaries, and seismic and volcanic activities overlap?*

6. Identify physical features at plate boundaries (move the mouse pointer over a physical feature to see the Map Tip).

(Q12) *Write the names of physical features in the first column of the table on the answer sheet and label them on the Supplement map. The first entry in the table is already completed for you as an example.*

7. In the table of contents, right-click plates and click Label Features.

You will make the text white and bold so the labels are easier to read.

B 8. Right-click plates and click Properties. Click the Labels tab. In the Text Symbol area, click the black color box and choose white. Click the Bold button. Click OK.

(Q13) *In the second column of the table on the answer sheet, write how you think each physical feature was created. Refer to the descriptions of the types of plate boundaries on pages 7 and 8.*

Step 9: Identify major cities at high or low risk for seismic or volcanic activity

1. Turn off earth_wsi.sid. Move the Major_cities layer to the top of the table of contents and turn it on.

2. Use the Zoom, Pan, and Identify tools to identify cities that have a high risk or low risk for an earthquake or a volcanic eruption.

(Q14) *List five high-risk cities and five low-risk cities. Remember to turn layers on and off and move them around as needed.*

3. Ask your teacher where to save this map document and how to rename it. If you are not going to save the document, choose Exit from the File menu. Click No when you are asked if you want to save changes.

In this lesson, you used different layers to find the locations of earthquakes and volcanoes around the world. You were then able to identify cities at high or low risk for seismic or volcanic events.

M2
L1

Name_____ Date_____

Answer sheet

Answer sheet
Module 2, Lesson 1

The earth moves

Step 2: Analyze earthquake locations

Q1) Do earthquakes occur in the places you predicted? List the regions you predicted correctly for earthquake locations.

_____

_____

_____

_____

Q2) What patterns do you see on the map?

_____

_____

_____

_____

Step 3: Sort and analyze earthquake magnitudes

Q3) How do the 20 selected locations compare to your Supplement map? List three ways.

_____

_____

_____

Step 4: Analyze volcano data

Q4) How do the volcano locations compare with your original predictions? List the regions of volcanic activity you predicted correctly.

_____

_____

_____

_____

_____

Q5) What patterns do you see in the volcano locations, and how do they compare with the earthquake patterns?

_____

_____

_____

_____

_____

Step 5: Select all active volcanoes

Q6) What pattern do you see?

_____

_____

_____

_____

_____

Q7) Formulate a hypothesis as to why volcano eruptions and earthquakes happen where they do.

_____

_____

_____

_____

_____

Step 6: Identify active volcanoes on different continents

Q8) Use the Identify tool to find the names, elevations, and countries of three active volcanoes.

_____

_____

_____

_____

_____

Step 7: Add the plate boundaries layer

Q9) Based on the locations of earthquakes and volcanoes, where do you think the plate boundaries are? Draw them on the Supplement map.

Q10) Compare the actual plate boundaries to the ones you drew on the Supplement map. Record all similarities and differences.

_____

_____

_____

_____

_____

_____

Step 8: Add a layer file and an image

Q11) Are there any areas where physical features, plate boundaries, and seismic and volcanic activities overlap?

_____

_____

_____

_____

_____

Q12) Write the names of physical features in the first column of the table below and label them on the Supplement map. The first entry is already completed for you as an example.

Physical feature	How the feature was created
Mid-Atlantic Ridge	Divergence of the South America and Africa plates and divergence of the North America and Eurasia plates

Q13) In the second column of the table on the previous page, write how you think each physical feature was created. Refer to the descriptions of the types of plate boundaries in the activity.

Step 9: Identify major cities at high or low risk for seismic or volcanic activity

Q14) List five high-risk cities and five low-risk cities.

High-risk cities	Low-risk cities

Middle school assessment
Module 2, Lesson 1

The earth moves

Create a risk map

On the Assessment map:
1. Mark all plate boundaries and label the plates.
2. Label five physical features associated with volcanoes or earthquakes.
3. From the following list, identify five cities at high risk for a volcanic or seismic disaster (use the Global2 map document and the Find tool as needed). Mark and label them on your map.

San Francisco, USA   Bombay, India   Jakarta, Indonesia
Manila, Philippines   Bogotá, Colombia   Addis Ababa, Ethiopia
Mexico City, Mexico   Rome, Italy   Seattle, USA
Houston, USA   Madrid, Spain   Tokyo, Japan
Hong Kong, China   Reykjavik, Iceland
Sydney, Australia   Cairo, Egypt

Analyze the map

On a separate piece of paper, write a paragraph for each of the following items:
1. Describe the relationships you see between tectonic plate boundaries and areas at high risk for volcanic or seismic activity.
2. Explain why you selected each city on your map.
3. Rank your five cities in the order of risk, with 1 being the highest risk, and explain why you ranked them that way.
4. Compare your Supplement map to your Assessment map.

The World

Assessment map

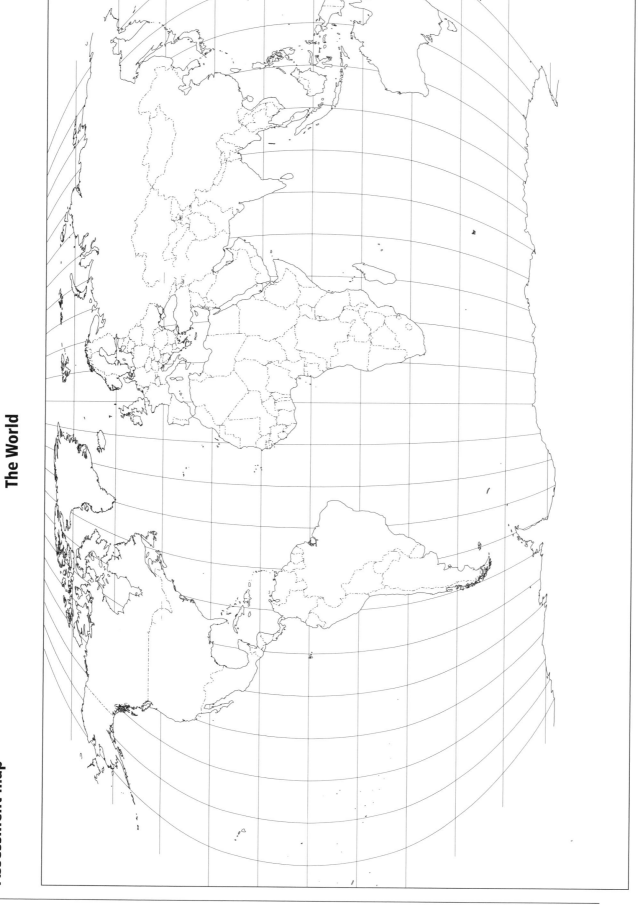

High school assessment
Module 2, Lesson 1

The earth moves

Create a risk map

On the Assessment map:
1. Mark all plate boundaries and label the plates.
2. Label five physical features associated with volcanoes or earthquakes.
3. From the following list, identify five cities at high risk for a volcanic or seismic disaster (use the Global2 map document and the Find tool as needed). Mark and label them on your map.

San Francisco, USA
Manila, Philippines
Mexico City, Mexico
Houston, USA
Hong Kong, China
Sydney, Australia

Bombay, India
Bogotá, Colombia
Rome, Italy
Madrid, Spain
Reykjavik, Iceland
Cairo, Egypt

Jakarta, Indonesia
Addis Ababa, Ethiopia
Seattle, USA
Tokyo, Japan

Analyze the map

On a separate piece of paper, write a paragraph for each of the following items:
1. Provide evidence that each of the cities you selected is at high risk for a major earthquake or volcanic eruption.
2. Choose at least two of the cities on your map and research how these cities are prepared (or not prepared) for a seismic disaster. For each city, write a complete paragraph outlining the city's preparedness.
3. Elaborate on your answers in Q13 as to how the five physical features you selected may have formed.

The World

Assessment map

Module 2, Lesson 2

Life on the edge

A regional investigation of East Asia

- Supplement: Map exercise
- Activity
- Answer sheet
- Assessment
- Assessment map A: East Asia
- Assessment map B: East Asia

Supplement

Map exercise

East Asia sits on the western edge of the Ring of Fire. This region experiences an enormous amount of geophysical activity, both seismic and volcanic. Use different colors to outline on the map below the areas you think have the greatest risk for volcanic and earthquake disasters (thick lines represent plate boundaries).

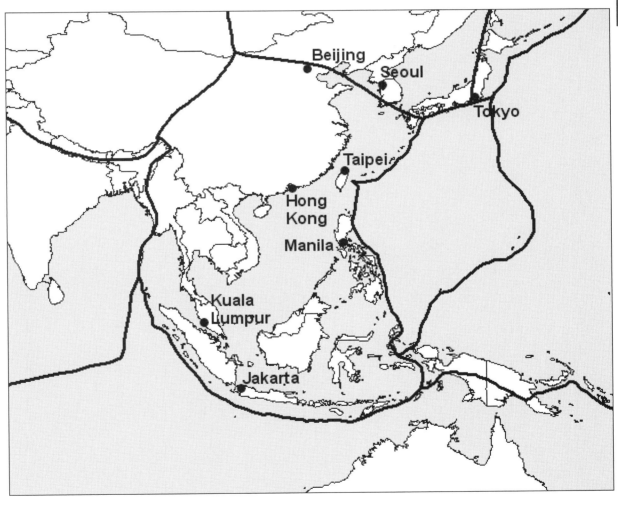

Module 2, Lesson 2

Life on the edge

In this activity, you will investigate the Pacific Ocean's Ring of Fire, with particular focus on earthquake and volcanic activity in East Asia, where millions of people live with the daily threat of significant seismic or volcanic events.

Step 1:   Open a map document and identify cities

1.   Double-click the ArcMap icon on your computer's desktop.

2.   When the ArcMap start-up dialog box appears, click **An existing map** and click OK.

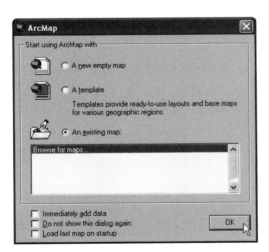

3. Navigate to the module 2 folder (**OurWorld2\Mod2**) and choose **Region2.mxd** (or **Region2**) from the list.

4. Click Open. When the map document opens, you see a map with two layers turned on: Major Cities and Countries. The check mark next to the layer name tells you the layer is turned on and visible in the map.

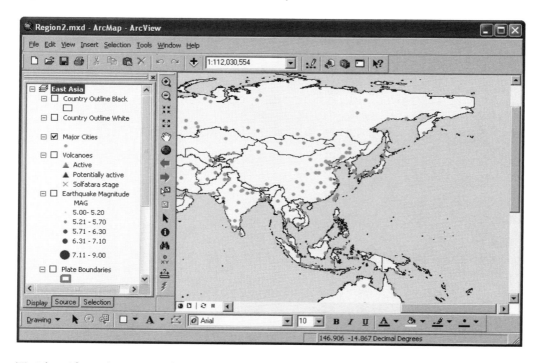

To identify each city on the map, you can use the Identify tool.

5. Click the Identify tool. An Identify window appears.

6. In the Identify window, click the "Identify from" drop-down menu and click Major Cities in the list. Move the cursor over the map. Notice how the cursor has a small "i" next to it.

7. Click on any green dot on the map.

The Identify dialog box appears similar to the graphic below:

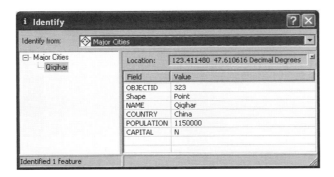

The right side of the Identify window tells you the name of the city, the country it's in, its population, and whether or not it's the capital city. In the example above, N = No (Qiqihar is not the capital of China).

Answers to questions in this activity should be recorded on the answer sheet.

Q1 *Use the Identify tool to locate one city in India and one in Japan. Record each city's name and population in the table on your answer sheet.*

8. Close the Identify window by clicking the X in the upper right corner of the window.

Step 2: Look at population density and earthquake magnitudes

1. Scroll down in the table of contents until you see Population Density. If necessary, widen the table of contents to see the Population Density legend.

2. Turn on the Population Density layer by clicking the box next to its name.

   A check mark appears, and the layer is drawn on the map. The darker areas on the map represent areas of high population density, as measured by the number of people per square kilometer.

Q2 *Use the Identify tool to locate two East Asian cities in areas where population density is greater than 250 people per square kilometer.*

3. Close the Identify window.

4. Turn off the Major Cities layer by clicking the box next to its name.

Q3 *Describe the general pattern of population density in East Asia. (To display a country's name, place your cursor over the country without clicking.)*

5. Turn on the Earthquake Magnitude layer.

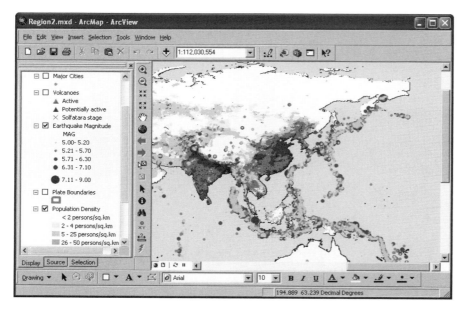

The purple dots indicate earthquake locations, with larger dots corresponding to stronger earthquakes. Only earthquakes with a magnitude of 5 or greater on the Richter scale are included in this layer.

Q4 *In general, where did earthquakes with a magnitude of ≥5 occur?*

Q5 *Did these earthquakes occur near densely populated areas? Where?*

Step 3: Measure the distance between active volcanoes and nearby cities

1. Turn off the Earthquake Magnitude layer.

2. Turn on the Volcanoes and Major Cities layers.

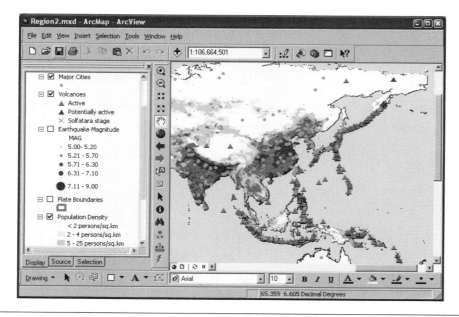

The volcanoes are grouped in three types: active, potentially active, and solfatara state (venting primarily hot gases). Note the different symbols used for each type of volcano.

3. Choose a city that's located near an active volcano. Click the Zoom In tool and then click the city's dot on the map. Your map is now centered on that city and zoomed in slightly.

4. Click the scale box above the map and choose 1:10,000,000. Your map is now zoomed in much closer to your city and the nearby volcano.

You will use the Measure tool to determine the distance between the volcano and the city.

5. Click the Measure tool. The Measure dialog box opens and your cursor turns into a right-angle ruler with crosshairs.

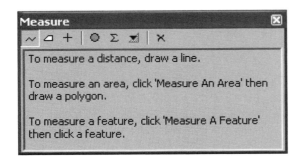

6. Click the Choose Units button and then click Distance and Kilometers.

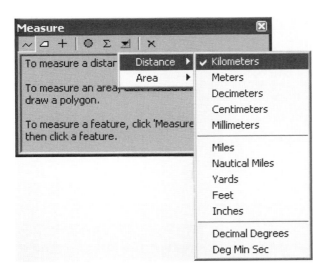

7. Click the volcano that is close to your chosen city. Now move your mouse pointer over to the city. A line is attached from the point where you first clicked to where you move your mouse pointer.

8. Once you have reached the city, double-click to end the line.

The distance in kilometers appears in the Measure dialog box. Line and segment distances are the same because the line contains only one segment.

> If you accidentally clicked the wrong spot, you can double-click to end the line and start over.

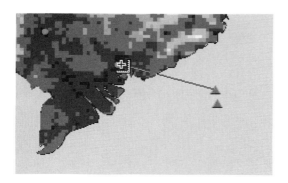

9. Use the Measure tool to determine the distance from other cities to nearby active volcanoes.

Q6 *What is the closest distance you found between a volcano and a city? Record that city, the volcano, and the distance between them. (Use the Identify tool as needed.)*

Q7 *Are there many active volcanoes located close to cities?*

Q8 *What patterns do you see in the locations of volcanoes, and how do they compare with the earthquake patterns? (Turn the Earthquake Magnitude layer on and off as needed.)*

10. At the top of the ArcMap window, click View, point to Bookmarks, and click East Asia. This takes you back to the view of East Asia.

Step 4:   Look at plate boundaries

1. Turn off all layers except Countries. Turn on the Plate Boundaries layer.

Now you will label the tectonic plates in the East Asia region.

2. In the table of contents, right-click Plate Boundaries and click Label Features. All the plates are now labeled on the map.

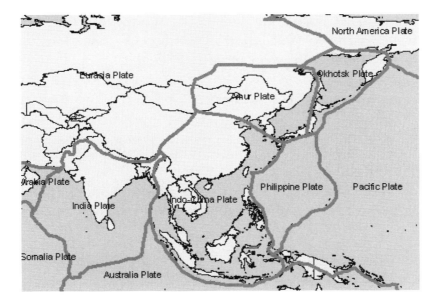

Ⓠ⁹ *Record the labels on the Supplement map.*

Many of the plate boundaries around the Pacific Rim have areas called subduction zones, where one plate is diving underneath another one. Subduction zones can be identified by underwater trenches and island arcs that are formed at these boundaries.

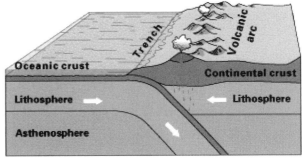

Oceanic-continental convergence

To get a closer look at physical features at plate boundaries, you will add the earth_wsi.sid image. This is a composite satellite image of the earth. It will allow you to see physical features in detail.

Step 5: Add an image file

1. Click the Add Data button and navigate to the Images folder (**OurWorld2\Mod2\ Data\Images**).

2. Select **earth_wsi.sid** and click Add to add it to your map.

   The image is displayed beneath the other layers because it is at the bottom of the table of contents.

☐ ☑ Plate Boundaries
☐ ☐ Population Density
    < 2 persons/sq.km
    2 - 4 persons/sq.km
    5 - 25 persons/sq.km
    26 - 50 persons/sq.km
    51 - 250 persons/sq.km
    251 - 1000 persons/sq.km
    1001 + persons/sq.km
☐ ☑ Countries
☐ ☑ earth_wsi.sid
    RGB
    Red: Band_1
    Green: Band_2
    Blue: Band_3

3. Find areas where subduction may be occurring. Remember, subduction zones are characterized by deep trenches and volcanic-island arcs and occur along plate boundaries. Turn the Plate Boundaries layer on and off as needed.

(Q10) *On the Supplement map, draw the zones of subduction.*

Step 6: Investigate your map

1. Use the Zoom and Pan tools to get a closer look at all the physical features. Refer to the ArcMap Toolbar Quick Reference for a brief explanation of the Zoom and Pan tools.

2. Use the Identify tool to find out the names of volcanoes or cities near plate boundaries. Turn layers on and off as needed.

    Remember, the map is drawn beginning with the layer listed at the bottom of the table of contents first and ending with the layer listed at the top. Move the layers around by clicking and dragging them up or down in the table of contents.

3. Explore one of the plate boundaries in detail, identifying cities and physical features in its vicinity.

(Q11) *Record the name of the plate you investigated. List three cities and three physical features in the vicinity.*

4. Ask your teacher how to save this map document and how to rename it. If you are not going to save the map document, choose Exit from the File menu. Click No when you are asked if you want to save changes.

    In this lesson, you used different layers to determine where earthquakes and volcanic eruptions are likely to occur in East Asia. You were then able to identify cities at risk.

Name_____   Date_____

Module 2, Lesson 2

Life on the edge

Step 1: Open a map document and identify cities

Q1)   Use the Identify tool to locate one city within each country listed in the table below and record that city's name and population.

City	Country	City population
Kunming	China	1,280,000
	India	
	Japan	

Step 2: Look at population density and earthquake magnitudes

Q2)   Use the Identify tool to locate two East Asian cities in areas where population density is greater than 250 people per square kilometer.

_____

_____

Q3)   Describe the general pattern of population density in East Asia.

_____

_____

_____

_____

_____

Q4) In general, where did earthquakes with a magnitude of ≥5 occur?

_____

_____

_____

_____

_____

Q5) Did these earthquakes occur near densely populated areas? Where?

_____

_____

_____

_____

_____

Step 3: Measure the distance between active volcanoes and nearby cities

Q6) What is the closest distance you found between a volcano and a city? Record that city, the volcano, and the distance between them.

City _____

Volcano _____

Distance _____

Q7) Are there many active volcanoes located close to cities? _____

Q8) What patterns do you see in the locations of volcanoes, and how do they compare with the earthquake patterns? (Turn the Earthquake Magnitude layer on and off as needed.)

_____

_____

_____

_____

_____

Step 4: Look at plate boundaries

Q9) Record the labels on the Supplement map.

Step 5: Add an image file

Q10) On the Supplement map, draw the zones of subduction.

Step 6: Investigate your map

Q11) Record the name of the plate you investigated. List three cities and three physical features in the vicinity.

Plate name: _____

Cities:

1. _____

2. _____

3. _____

Physical features:

1. _____

2. _____

3. _____

Middle school assessment
Module 2, Lesson 2

Life on the edge

Create a risk map

Use the map document from the activity and your class notes to do the following on the Assessment A map:

1. Mark zones at high risk for volcanic activity with one color and zones at low risk with another color. Take into account the following factors:
   - Population density
   - Proximity of active volcanoes to major cities
2. Mark zones at high risk for seismic activity with one color and zones at low risk with another color. Take into account the following factors:
   - Population density
   - Proximity of recent earthquakes to major cities
3. Create a map legend that identifies the four zones.

Analyze the map

On a separate piece of paper, write a paragraph for each of the following items:

1. Describe the criteria you used to define zones of risk for volcanic and earthquake activities.
2. Describe the relationships you see between tectonic plate boundaries and areas at high risk for volcanic and seismic activities.
3. Describe physical features around the Pacific Rim and identify their common characteristics.
4. Develop an emergency action plan for a city in one of the high-risk zones.

Assessment map A: East Asia

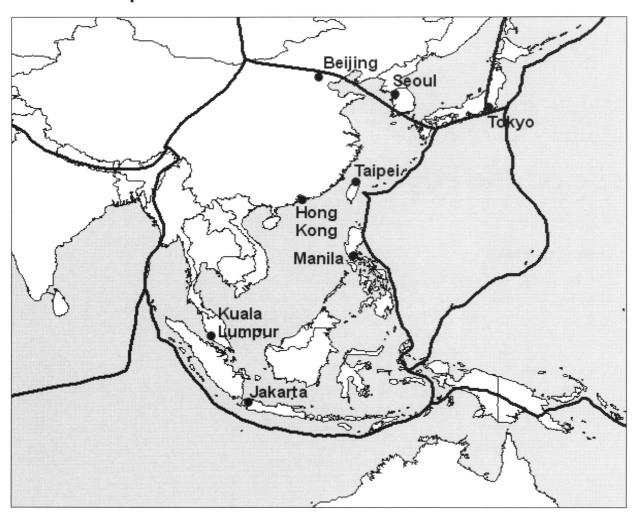

High school assessment
Module 2, Lesson 2
Life on the edge

Create a risk map

Use the map document from the lesson and your class notes to do the following on the Assessment B map:

1. Mark zones of high, medium, and low risk for volcanic activity with different colors. Take into account the following factors:
   - Population density
   - Proximity of active volcanoes to major population centers
   - Locations of plate boundaries and subduction zones
2. Mark zones of high, medium, and low risk for seismic activity with different colors. Take into account the following factors:
   - Population density
   - Magnitude of earthquakes from 2004 to 2007 data
   - Locations of plate boundaries and subduction zones
3. Create a complete map legend.
4. Mark and label major cities in the high-risk zones (use the Find tool as needed), plate boundaries, and subduction zones.

Analyze the map

On a separate piece of paper, write a paragraph for each of the following items:

1. Describe the criteria you used to define zones of risk for volcanic and earthquake activities.
2. Describe the relationships you observe between tectonic plate boundaries and areas at high risk for volcanic and seismic activities.
3. Explain how subduction zones affect three different Pacific Rim physical features.
4. Develop an emergency action plan for a city in one of the high-risk zones.

Assessment map B: East Asia

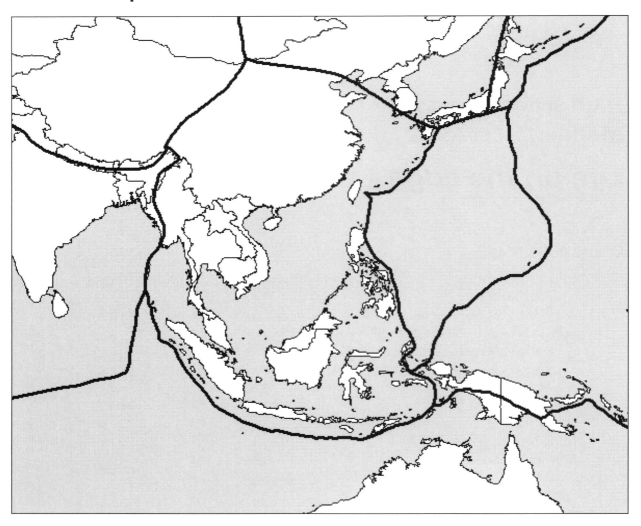

Module 3, Lesson 1

Running hot and cold

A global perspective

- Supplement: Hot and cold cities
- Activity
- Answer sheet
- Assessment

Supplement

Hot and cold cities

Of the cities on the map below, list three that you believe are the hottest in July and three that you believe are the coldest in January.

Hottest in July: Coldest in January:

_____   _____

_____   _____

_____   _____

Module 3, Lesson 1

Running hot and cold

In this activity, you will analyze monthly and annual temperature patterns in cities around the world. You will explore how latitude, elevation, and proximity to the ocean influence temperature patterns in the world's tropical, temperate, and polar zones.

Step 1: Open a map document

1. Double-click the ArcMap icon on your computer's desktop.

2. When the ArcMap start-up dialog box appears, click **An existing map** and click OK.

3. Navigate to the module 3 folder (**OurWorld2\Mod3**) and choose **Global3.mxd** (or **Global3**) from the list.

4. Click Open. When the map document opens, you see a world map.

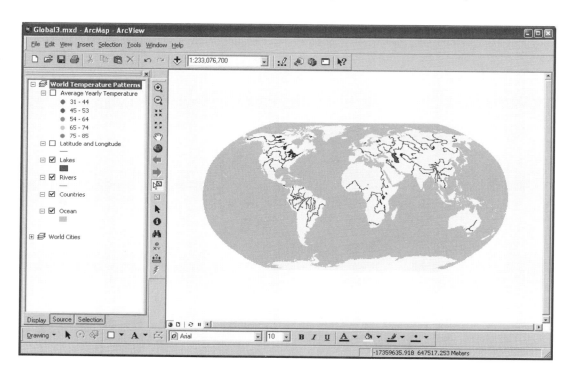

Step 2:   Observe annual world temperature patterns

1. Turn on the Average Yearly Temperature layer.

   The symbols on the map represent cities around the world. The color of each symbol reflects an average of temperatures recorded throughout the year in that city (in degrees Celsius).

2. Look at the global temperature patterns displayed on the map.

Q1 *Write three observations about the pattern of temperatures displayed on the map. Your observations should describe regions of the world, not specific countries or cities.*

3. Click the check mark next to the Average Yearly Temperature layer to turn it off.

Step 3: Label latitude zones

1. Turn on the Latitude and Longitude layer.

2. In the table of contents, right-click Latitude and Longitude and click Properties. Click the Labels tab.

3. At the top of the Labels tab, click the check box next to "Label features in this layer." Notice that NAME is already chosen as the field to use for labeling.

4. In the Text Symbol section, use the drop-down menus to set the font to Arial and the size to 9. Set the style to Bold.

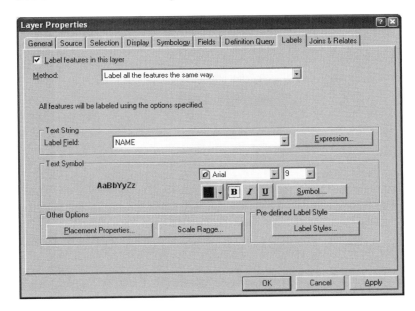

5. Click OK. The major latitude and longitude lines are labeled on the map.

In order to display the label for the prime meridian, you may need to zoom in or enlarge your map document window.

For the purposes of this exercise, the areas between the major latitude lines represent five zones of latitude. The table below names each latitude zone and the area it covers.

Zone	Latitude range
North Polar	Arctic Circle–North Pole
North Temperate	Tropic of Cancer–Arctic Circle
Tropical	Tropic of Cancer–Tropic of Capricorn
South Temperate	Tropic of Capricorn–Antarctic Circle
South Polar	Antarctic Circle–South Pole

Now you will label each of these zones on your map.

6. On the Draw toolbar at the bottom of the ArcMap window, use the drop-down menus to set the font to Arial, the size to 9, and the style to Bold.

If you don't see the Draw toolbar, right-click in the gray area near the top of the ArcMap window and click Draw to turn it on.

7. Click the Font Color drop-down menu and choose a deep violet color.

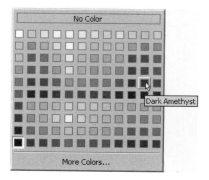

8. Zoom in to the area between the Arctic Circle and the North Pole.

9. On the Draw toolbar, click the New Text tool. The cursor turns into a plus sign with the letter A when you move it over the map.

10. Click the map somewhere above the Arctic Circle. A text box appears.

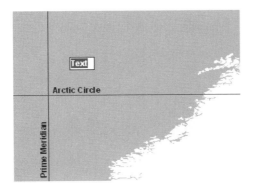

11. Type North Polar Zone in the box and press Enter. The North Polar Zone is now labeled in violet on the map.

> If you make a spelling mistake, double-click the text. Correct the mistake in the dialog box and click OK.

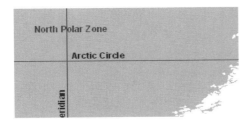

12. Label each of the remaining zones using the same procedure you used to label the North Polar Zone. Remember, the zones are the spaces between the lines of latitude. You may want to label the Tropical Zone twice, north and south of the equator.

    You can use the scroll bar to the right of the map display or the Pan tool to move up or down on your map.

13. Click the Full Extent button to see the whole world.

14. In the table of contents, right-click Latitude and Longitude and click Label Features to turn off the latitude and longitude (black) labels. Now you are left with the new latitude zones (violet) labels.

    You can toggle labels on and off by right-clicking a layer and clicking Label Features. A check mark on the menu next to Label Features indicates that labels are turned on.

M3
L1

15. Click the Select Elements tool. Click any label on the map that you want to move, then click it again and drag it. Move the labels to the ocean area so that they cover very little of the continents.

16. Turn on the Average Yearly Temperature layer. Observe the temperature patterns as they correspond to the latitude zones.

17. Click the Identify tool. The Identify window opens.

18. Click the "Identify from" drop-down menu and choose Average Yearly Temperature.

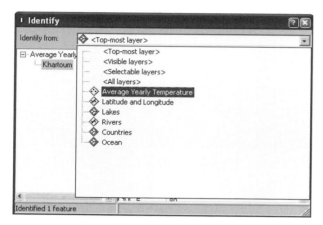

Q2 Use the Identify tool to click cities and get the necessary information to complete the table on your answer sheet (remember that each dot represents a city).

Q3 Why do you think there aren't any major cities in the North or South Polar Zones?

Q4 How is the North Temperate Zone different from the South Temperate Zone?

19. Turn off Average Yearly Temperature and close the Identify window.

Step 4: Observe climate distribution

1. Click the Add Data button.

2. Navigate to the module 3 LayerFiles folder (**OurWorld2\Mod3\Data\LayerFiles**). Select **Climate.lyr** and click Add.

The Climate layer displays the regions of the world characterized by different types of climate.

Q5 Complete the table on your answer sheet.

Q6 Which zone has the greatest number of climates?

3. Turn on the Average Yearly Temperature layer.

> **Q7** *Give an example of a city in each of the climate zones listed in the answer sheet (use the Identify tool to get the names of the cities).*

> **Q8** *Ask your teacher if you should stop now and save your work. If so, ask for instructions on how to rename and save this map document. Record the new name of the map document and its new location on your answer sheet. Then from the File menu, click Exit.*

Step 5: Observe monthly temperature patterns in the Northern Hemisphere

1. If necessary, start ArcMap and locate the map document where you saved it in Q8. Refer to your answer sheet for its name and location. Open the map document.

2. In the table of contents, click the minus sign in front of the World Temperature Patterns data frame to collapse it.

3. Click the plus sign in front of the World Cities data frame to expand it. Right-click World Cities and click Activate.

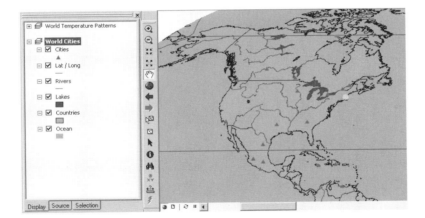

You see a map centered on North America showing cities, rivers, and lakes. The symbol for Boston is highlighted on the map. If your map looks different than the one pictured above, use the Zoom In tool to drag a box around North America.

4. Click the Tools menu, point to Graphs, and click Monthly Temperature.

The graph displays the monthly temperatures for Boston. You're going to select additional cities on the map, but first you'll reposition the graph and set the selectable layer to Cities.

5. Click the graph's title bar and drag it so that it doesn't obstruct the map.

6. At the bottom of the table of contents, click the Selection tab. Notice that one city (Boston) is selected.

Display | Source | Selection

7. Click the check boxes for all layers except Cities to uncheck them.

☑ **Cities (1)**
☐ Lat / Long
☐ Rivers
☐ Lakes
☐ Countries
☐ Ocean

8. Click the Display tab to show the table of contents again.

9. Click the Select Features tool.

10. Click on Miami at the southern tip of Florida (if you are not sure which city is Miami, point to a city without clicking to display its name).

> If you get a graph with many colored lines, click directly on the city symbol again to get a single line graph.

Notice that both the map and the graph have changed.

(Q9) *What does the graph show now?*

(Q10) *What city is highlighted on the map?*

11. Hold down the Shift key and select Boston again on the map.

(Q11) *What does the graph show now?*

(Q12) *What city or cities are highlighted on the map?*

(Q13) *Use the Monthly Temperature graph to compare Miami and Boston and complete the table on your answer sheet.*

12. Hold down the Shift key and click the city northeast of Boston.

(Q14) *What is the name of the city?*

Q15 *How does its monthly temperature pattern differ from Boston's?*

13. Hold down the Shift key and click the closest city south of Miami.

Q16 *What is the name of the city?*

Q17 *How does its monthly temperature pattern differ from Miami's?*

14. Pause your cursor over a city symbol on the map to find that city's coordinates. The coordinates (longitude followed by latitude) are displayed on the status bar at the bottom of the ArcMap window.

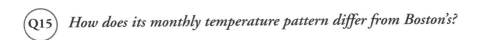

-72.106 47.368 Decimal Degrees

The coordinates are displayed in decimal degrees rather than in degrees, minutes, and seconds. Latitudes north of the equator and longitudes east of the prime meridian are positive numbers, whereas latitudes south of the equator and longitudes west of the prime meridian are negative numbers.

Q18 *List the name of each of the cities displayed in the Monthly Temperature graph and complete the information in the table on the answer sheet.*

Q19 *Based on the information displayed in the graph, the map, and the table in Q18, formulate a hypothesis about how the monthly temperature patterns change as latitude increases.*

Step 6:   Test your hypothesis

1. Click the Clear Selected Features button.

   The Monthly Temperature graph now shows many lines. ArcMap graphs the temperatures for all of the cities in the layer when no city is selected.

2. Click the Pan tool. Click the map and pan over to Western Europe.

3. Click the Find tool. The Find dialog box displays.

4. Type **Stockholm** in the Find text box and click Find. A record (row) displays in the white box at the bottom of the dialog box.

5. Right-click the row and click Select. Stockholm is highlighted in blue on the map.

6. Close the Find dialog box.

7. Click the Select Features tool. Hold down the Shift key and select three more European cities that are increasingly south of Stockholm. The city names appear in the graph and the cities are highlighted on the map.

> To unselect a city that you selected by mistake, hold down the Shift key and click that city. To make the legend on the graph more readable, enlarge the graph window by dragging any of its borders.

(Q20) *Complete the table on the answer sheet.*

(Q21) *Does the data for the cities you selected confirm or dispute your hypothesis in Q19? Explain.*

Step 7: Analyze temperature patterns in the Southern Hemisphere

You've already formulated a hypothesis about how latitude affects monthly temperature patterns in the Northern Hemisphere. Now you will explore the effect of latitude on the monthly temperature patterns within the Southern Hemisphere.

 1. Clear the selected features.

2. Click the Pan tool and reposition the map so that it is centered on Australia.

3. Click the Find tool to locate and select the city of Darwin as you did for Stockholm above.

4. Close the Find dialog box.

5. Click the Select Features tool. Hold down the Shift key and click the three cities on Australia's eastern and southern coasts.

(Q22) *Complete the table on your answer sheet.*

(Q23) *Based on the information displayed on the graph, the map, and the table you just completed, compare the monthly temperature patterns in the Southern Hemisphere to those in the Northern Hemisphere.*

(Q24) *Formulate a hypothesis about the relationship between monthly temperature patterns and increases in latitude in the Southern Hemisphere.*

Step 8: Test your hypothesis for the Southern Hemisphere

1. Clear the selected features.

2. Click the Pan tool and reposition the map so that it is centered on and includes all of Africa (zoom in or out if you need to).

3. Use the Find tool to locate and select the city of Cape Town as you did for Stockholm above.

4. Close the Find dialog box.

5. Select two or three more African cities located between Cape Town and the equator.

> (Q25) *Complete the table on your answer sheet.*
>
> (Q26) *Does the data for the cities you selected confirm or dispute your hypothesis about how latitude affects monthly temperature patterns in the Southern Hemisphere? Explain.*

6. Clear the selected features.

7. Reposition your map so that it's centered on North America.

Step 9: Investigate the ocean's influence on temperature

In addition to latitude and hemisphere, a city's proximity to the ocean also influences its temperature. Now you will investigate how the ocean influences the air temperature of coastal cities.

1. Make sure that World Cities is the active data frame.

2. One at a time, select all the cities in Canada.

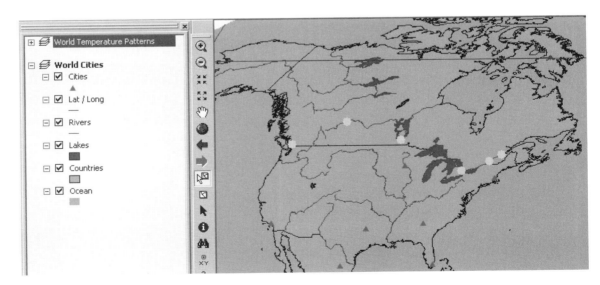

(Q27) *In which Canadian city would you experience the coldest winter temperatures?*

(Q28) *In which Canadian city would you experience the warmest winter temperatures?*

(Q29) *Looking at the map, why do you think the warmest city has winter temperatures that are so much warmer than the others? (Think of how this city is different from all the others in terms of its location.)*

3. Clear the selected features.

4. Reposition your map so that it is centered on Western Europe.

5. In the table of contents, right-click Cities and click Open Attribute Table.

6. Click the title bar for the Attributes of Cities table and move it so that you can still see all of Western Europe on the map (use your mouse to resize the table if it is too big).

7. Click the column heading NAME. The heading looks like a button that's been pushed in, and the column is highlighted in blue.

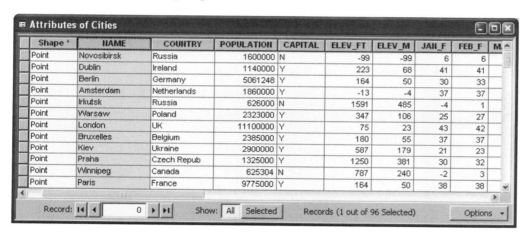

Shape *	NAME	COUNTRY	POPULATION	CAPITAL	ELEV_FT	ELEV_M	JAN_F	FEB_F	M
Point	Novosibirsk	Russia	1600000	N	-99	-99	6	6	
Point	Dublin	Ireland	1140000	Y	223	68	41	41	
Point	Berlin	Germany	5061248	Y	164	50	30	33	
Point	Amsterdam	Netherlands	1860000	Y	-13	-4	37	37	
Point	Irkutsk	Russia	626000	N	1591	485	-4	1	
Point	Warsaw	Poland	2323000	Y	347	106	25	27	
Point	London	UK	11100000	Y	75	23	43	42	
Point	Bruxelles	Belgium	2385000	Y	180	55	37	37	
Point	Kiev	Ukraine	2900000	Y	587	179	21	23	
Point	Praha	Czech Repub	1325000	Y	1250	381	30	32	
Point	Winnipeg	Canada	625304	N	787	240	-2	3	
Point	Paris	France	9775000	Y	164	50	38	38	

Record: 0   Show: All Selected   Records (1 out of 96 Selected)   Options ▾

8. Right-click the column heading and click Sort Ascending to make the list alphabetical. Select Amsterdam in the list by clicking the small gray box in the first column of that row. The row that Amsterdam is in is highlighted in blue.

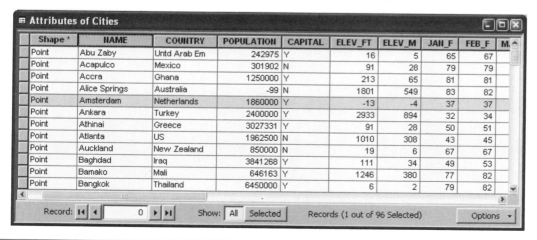

Shape *	NAME	COUNTRY	POPULATION	CAPITAL	ELEV_FT	ELEV_M	JAN_F	FEB_F	M
Point	Abu Zaby	Untd Arab Em	242975	Y	16	5	65	67	
Point	Acapulco	Mexico	301902	N	91	28	79	79	
Point	Accra	Ghana	1250000	Y	213	65	81	81	
Point	Alice Springs	Australia	-99	N	1801	549	83	82	
Point	Amsterdam	Netherlands	1860000	Y	-13	-4	37	37	
Point	Ankara	Turkey	2400000	Y	2933	894	32	34	
Point	Athinai	Greece	3027331	Y	91	28	50	51	
Point	Atlanta	US	1962500	N	1010	308	43	45	
Point	Auckland	New Zealand	850000	N	19	6	67	67	
Point	Baghdad	Iraq	3841268	Y	111	34	49	53	
Point	Bamako	Mali	646163	Y	1246	380	77	82	
Point	Bangkok	Thailand	6450000	Y	6	2	79	82	

Record: 0   Show: All Selected   Records (1 out of 96 Selected)   Options ▾

9. Scroll down the list. Hold down the Ctrl key and select the following cities, taking note of where each one is on the map as you select it: Berlin, Kiev, London, and Warsaw.

All of the cities are highlighted in the table and on the map, and they are displayed on the graph.

10. Move the table out of the way or minimize it if necessary so that you can see the map and graph. Analyze the map and graph.

(Q30) *Complete the table on your answer sheet.*

(Q31) *What do these cities have in common in terms of their locations on the earth?*

(Q32) *Which two cities have the mildest temperatures?*

(Q33) *What happens to the winter temperatures as you move from London to Kiev?*

(Q34) *Why do you think some cities have milder temperatures than the others?*

(Q35) *Based on your observations for Canada (Q27–Q29) and Western Europe, formulate a hypothesis about the influence of proximity to the ocean (or distance from it) on patterns of temperature.*

Step 10: Investigate the impact of elevation on temperature patterns

A city's elevation significantly affects temperature in that city. You will now investigate the relationship between elevation and temperature.

1. Click the Full extent button.

2. Restore the Attributes of Cities table if you minimized it.

3. Scroll up in the table until you find the city of Kisangani. Click the small gray box to the left of this record to select it.

4. Scroll down in the table and, holding down the Ctrl key, select Libreville, Quito, and Singapore. Take note of where each one is on the map as you select it in the table (move the table out of the way or minimize it so that you can see the map and the graph).

(Q36) *Complete the table on your answer sheet.*

(Q37) *What do these cities have in common in terms of their locations on the earth?*

(Q38) *What temperature pattern do these four cities have in common?*

Q39 *How is Quito different from the other three?*

Q40 *Since all these cities are located on or very near the equator, what other factor could explain the difference in their temperature patterns?*

5. Restore the Attributes of Cities table if necessary.

6. Click the Selected button at the bottom of the table. Only the selected records are displayed in the table.

7. Right-click on the NAME column heading and click Freeze/Unfreeze Column. Scroll to the right until the ELEV_M (elevation in meters) field is adjacent to the NAME field.

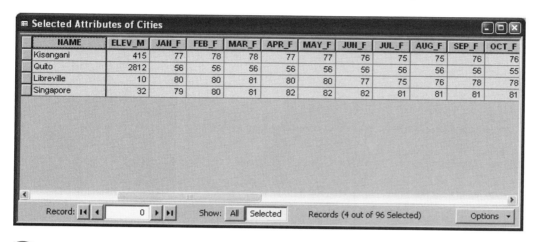

NAME	ELEV_M	JAN_F	FEB_F	MAR_F	APR_F	MAY_F	JUN_F	JUL_F	AUG_F	SEP_F	OCT_F
Kisangani	415	77	78	78	77	77	76	75	75	76	76
Quito	2812	56	56	56	56	56	56	56	56	56	55
Libreville	10	80	80	81	80	80	77	75	76	78	78
Singapore	32	79	80	81	82	82	82	81	81	81	81

Record: [|◄] [◄] 0 [►] [►|] Show: All | Selected   Records (4 out of 96 Selected)   Options ▼

Q41 *Analyze the selected records and complete the table on your answer sheet.*

8. Minimize the Attributes of Cities table.

9. Compare the elevation table on your answer sheet to the graph on your computer.

Q42 *Based on your observation of temperatures along the equator and the information in the table in Q41, formulate a hypothesis about the influence of elevation on patterns of temperature.*

10. Clear selected features.

Step 11: Revisit your initial ideas

1. Take out the paper map you used at the beginning of the lesson to identify the three coldest cities in January and the three hottest cities in July.

2. Restore the Attributes of Cities table and select the 13 cities that appear on your paper map.

3. Click the Selected button.

4. Scroll to the right until you see the column heading JAN_C (January temperature in Celsius). The NAME field should still be frozen.

5. Click the JAN_C column heading to select it. The column is highlighted yellow.

6. Right-click the JAN_C column heading and click Sort Ascending. The table is sorted from lowest (coldest) to highest (hottest) January temperatures.

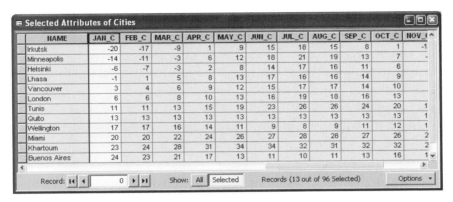

NAME	JAN_C	FEB_C	MAR_C	APR_C	MAY_C	JUN_C	JUL_C	AUG_C	SEP_C	OCT_C	NOV_(
Irkutsk	-20	-17	-9	1	9	15	18	15	8	1	-1
Minneapolis	-14	-11	-3	6	12	18	21	19	13	7	-
Helsinki	-6	-7	-3	2	8	14	17	16	11	6	
Lhasa	-1	1	5	8	13	17	16	16	14	9	
Vancouver	3	4	6	9	12	15	17	17	14	10	
London	6	6	8	10	13	16	19	18	16	13	
Tunis	11	11	13	15	19	23	26	26	24	20	1
Quito	13	13	13	13	13	13	13	13	13	13	1
Wellington	17	17	16	14	11	9	8	9	11	12	1
Miami	20	20	22	24	26	27	28	28	27	26	2
Khartoum	23	24	28	31	34	34	32	31	32	32	2
Buenos Aires	24	23	21	17	13	11	10	11	13	16	1

Record: 0   Show: All Selected   Records (13 out of 96 Selected)   Options

Q43 *On your answer sheet, rank the 13 cities from coldest to hottest according to their average January temperatures.*

7. Scroll to the right in the table and click the column heading JUL_C (July temperatures in Celsius).

8. Right-click the JUL_C column heading and click Sort Descending. The table is sorted from highest to lowest July temperatures.

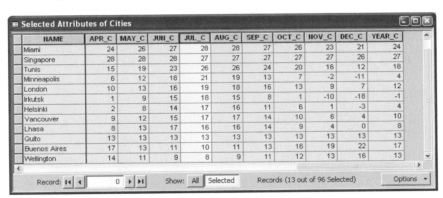

NAME	APR_C	MAY_C	JUN_C	JUL_C	AUG_C	SEP_C	OCT_C	NOV_C	DEC_C	YEAR_C
Miami	24	26	27	28	28	27	26	23	21	24
Singapore	28	28	28	27	27	27	27	27	26	27
Tunis	15	19	23	26	26	24	20	16	12	18
Minneapolis	6	12	18	21	19	13	7	-2	-11	4
London	10	13	16	19	18	16	13	9	7	12
Irkutsk	1	9	15	18	15	8	1	-10	-18	-1
Helsinki	2	8	14	17	16	11	6	1	-3	4
Vancouver	9	12	15	17	17	14	10	6	4	10
Lhasa	8	13	17	16	16	14	9	4	0	8
Quito	13	13	13	13	13	13	13	13	13	13
Buenos Aires	17	13	11	10	11	13	16	19	22	17
Wellington	14	11	9	8	9	11	12	13	16	13

Record: 0   Show: All Selected   Records (13 out of 96 Selected)   Options

Q44 *Rank the 13 cities from hottest to coldest according to their average July temperatures.*

9. Compare your original predictions from your paper map with the correct answers on your answer sheet.

Q45 *Put a check mark next to the answers in Q43 and Q44 that you predicted correctly.*

10. At the bottom of the attribute table, click the All button.

M3
L1

11. Also at the bottom of the attribute table, click the Options button and click Clear Selection on the menu.

12. Close the attribute table and the graph. (There are no cities selected on the map at this time.)

13. Save your changes.

14. Exit ArcMap by choosing Exit from the File menu.

In this lesson you explored temperature data for 96 world cities. You explored the different latitude zones and have identified the variety of climates in each zone. You formulated and tested different hypotheses to explain temperature patterns. Now you know how latitude, hemisphere, proximity to the ocean, and elevation affect temperature patterns around the world.

Name_____ Date_____

Module 3, Lesson 1

Running hot and cold

Step 2: Observe annual world temperature patterns

Q1) Write three observations about the pattern of temperatures displayed on the map. Your observations should describe regions of the world, not specific countries or cities.

_____

_____

_____

Step 3: Label latitude zones

Q2) Complete the following table.

Zone	Typical temperature range (C)	Typical city	Atypical cities
Tropical			
North Temperate			
South Temperate			

Q3) Why do you think there aren't any major cities in the North or South Polar Zones?

_____

_____

_____

Q4) How is the North Temperate Zone different from the South Temperate Zone?

_____

_____

_____

Step 4: Observe climate distribution

Q5) Complete the following table.

Zone(s)	Characteristic climate(s)
Tropical	
Temperate	
Polar	

Q6) Which zone has the greatest number of climates? _____

Q7) Give an example of a city in each of the following climate zones:

Arid _____

Tropical Wet _____

Tropical Dry _____

Humid Subtropical _____

Mediterranean _____

Marine _____

Humid Continental _____

Subarctic _____

Highland _____

Q8)   Record the new name of the map document and its new location.

Document _____
    (Example: ABC_Mod3Les1.mxd)

Location _____
    (Example: C:\Student\ABC)

Step 5: Observe monthly temperature patterns in the Northern Hemisphere

Q9)   What does the graph show now? _____

Q10)   What city is highlighted on the map? _____

Q11)   What does the graph show now? _____

Q12)   What city or cities are highlighted on the map? _____

Q13)   Complete the following table.

City	Coldest months	Lowest temperature (℃)	Hottest months	Highest temperature (℃)	Temperature range over 12 months (℃)
Boston					
Miami					

Q14)   What is the name of the city? _____

Q15)   How does its monthly temperature pattern differ from Boston's?

_____

_____

Q16)   What is the name of the city? _____

Q17) How does its monthly temperature pattern differ from Miami's?

_____

_____

Q18) Complete the following table for each of the cities displayed in the graph.

City	Latitude	Coldest month(s)	Lowest temperature (C)	Hottest month(s)	Highest temperature (C)	Temperature range over 12 months
Boston						
Miami						

Q19) Based on the information displayed in the graph, the map, and the table in Q18, formulate a hypothesis about how the monthly temperature patterns change as latitude increases.

_____

_____

Step 6: Test your hypothesis

Q20) Complete the following table.

City	Latitude
Stockholm	

Q21) Does the data for the cities you selected confirm or dispute your hypothesis in Q19?

Explain. _____

_____

_____

Step 7: Analyze temperature patterns in the Southern Hemisphere

Q22) Complete the following table.

City	Latitude (°)	Coldest month(s)	Lowest temperature (C)	Hottest month(s)	Highest temperature (C)	Temperature range over 12 months
Darwin						

Q23) Compare the monthly temperature patterns in the Southern Hemisphere to those in the Northern Hemisphere.

_____

_____

_____

Q24) Formulate a hypothesis about the relationship between monthly temperature patterns and increases in latitude in the Southern Hemisphere.

_____

_____

_____

Step 8: Test your hypothesis for the Southern Hemisphere

Q25) Complete the following table.

City	Latitude (˚)
Cape Town	

Q26) Does the data for the cities you selected confirm or dispute your hypothesis about how latitude affects monthly temperature patterns in the Southern Hemisphere?

Explain. _____

_____

_____

_____

Step 9: Investigate the ocean's influence on temperature

Q27) In which Canadian city would you experience the coldest winter temperatures?

_____

Q28) In which Canadian city would you experience the warmest winter temperatures?

_____

Q29) Looking at the map, why do you think the warmest city has winter temperatures that are so much warmer than the others?

_____

_____

_____

_____

_____

Q30) Complete the following table.

City	Latitude (°)
London	
Amsterdam	
Berlin	
Warsaw	
Kiev	

Q31) What do these cities have in common in terms of their locations on the earth?

_____

Q32) Which two cities have the mildest temperatures?

_____

_____

Q33) What happens to the winter temperatures as you move from London to Kiev?

_____

Q34) Why do you think some cities have milder temperatures than the others?

_____

Q35) Based on your observations for Canada (Q27–Q29) and Western Europe, formulate a hypothesis about the influence of proximity to the ocean (or distance from it) on patterns of temperature.

_____

_____

_____

_____

Step 10: Investigate the impact of elevation on temperature patterns

Q36) Complete the following table.

City	Latitude (°)
Kisangani	
Libreville	
Quito	
Singapore	

Q37) What do these cities have in common in terms of their locations on the earth?

_____

Q38) What temperature pattern do these four cities have in common?

_____

Q39) How is Quito different from the other three?

_____

_____

Q40) Since all these cities are located on or very near the equator, what other factor could explain the difference in their temperature patterns?

_____

_____

Q41) Analyze the selected records and complete the following table.

City	Elevation (meters)
Kisangani	
Libreville	
Quito	
Singapore	

Q42) Based on your observation of temperatures along the equator and the information in the table in Q41, formulate a hypothesis about the influence of elevation on patterns of temperature.

_____

_____

_____

_____

Step 11: Revisit your initial ideas

Q43) Rank the 13 cities from coldest to hottest according to their average January temperatures.

1.	8.
2.	9.
3.	10.
4.	11.
5.	12.
6.	13.
7.	

Q44) Rank the 13 cities from hottest to coldest according to their average July temperatures.

1.	8.
2.	9.
3.	10.
4.	11.
5.	12.
6.	13.
7.	

Q45) Put a check mark next to the answers in Q43 and Q44 that you predicted correctly.

Middle school assessment
Module 3, Lesson 1

Running hot and cold

1. Use the ArcMap map document and your answer sheet to complete the tables below. For each city, circle each factor that influences its temperature pattern.

Three hottest cities in July	Factor(s) that influence temperature patterns		
	Latitude	Proximity to ocean	Elevation
	Latitude	Proximity to ocean	Elevation
	Latitude	Proximity to ocean	Elevation

Three hottest cities in January	Factor(s) that influence temperature patterns		
	Latitude	Proximity to ocean	Elevation
	Latitude	Proximity to ocean	Elevation
	Latitude	Proximity to ocean	Elevation

2. Use the information from the lesson, the Global3 map document, and other resources such as an atlas to write an essay that compares monthly and annual temperature patterns typical of the Tropical Zone and the North and South Temperate zones. Your essay should provide example cities and data to support your conclusions.

Create a paper map or a GIS-generated map that illustrates the conclusions you make in your essay.

High school assessment
Module 3, Lesson 1

Running hot and cold

1. Use the ArcMap map document and your answer sheet to complete the tables below. For each city, circle each factor that influences its temperature pattern.

Three hottest cities in July	Factor(s) that influence temperature patterns		
	Latitude	Proximity to ocean	Elevation
	Latitude	Proximity to ocean	Elevation
	Latitude	Proximity to ocean	Elevation

Three hottest cities in January	Factor(s) that influence temperature patterns		
	Latitude	Proximity to ocean	Elevation
	Latitude	Proximity to ocean	Elevation
	Latitude	Proximity to ocean	Elevation

2. Use the information from the lesson, the Global3 map document, and other resources such as an atlas to write an essay that compares monthly and annual temperature patterns typical of the Tropical Zone and the North and South Temperate zones. Your essay should provide example cities and data to support your conclusions.

Create a GIS-generated map that illustrates the conclusions you make in your essay.

Module 3, Lesson 2

Seasonal differences

A regional investigation of South Asia

- Activity
- Answer sheet
- Assessment

Module 3, Lesson 2

Seasonal differences

In this activity, you will analyze the variable patterns of precipitation in South Asia that result from the region's seasonal monsoon winds. As you investigate those patterns, you will explore relationships between rainfall and physical features and analyze the climate's impact on agriculture and population.

Step 1: Open a map document

1. Double-click the ArcMap icon on your computer's desktop.

2. When the ArcMap start-up dialog box appears, click **An existing map** and click OK.

3. Navigate to the module 3 folder (**OurWorld2\Mod3**) and choose **Region3.mxd** (or **Region3**) from the list. When the map document opens, you see a map of South Asia.

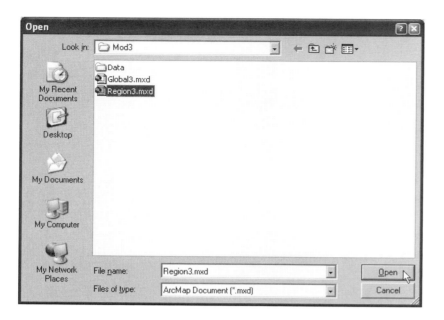

4. Click the Tools menu, point to Graphs, and click Monthly Rainfall. A graph of monthly rainfall for the city of Mumbai opens.

5. Click the graph's title bar and position it anywhere on your desktop that does not cover your map. Stretch or shrink the ArcMap window if you need to.

6. Repeat step 4 to open the Annual Rainfall graph. Position it next to the Monthly Rainfall graph.

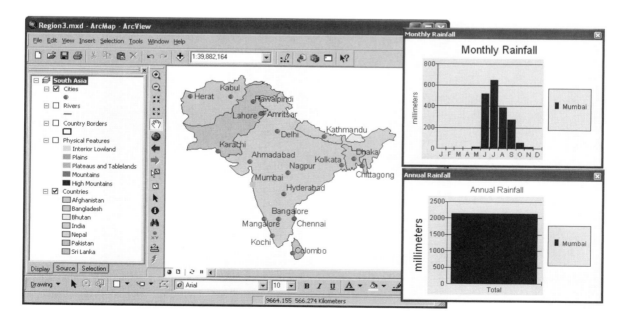

Step 2: Observe patterns of rainfall

The map allows you to explore and compare variations in the patterns of rainfall throughout the South Asian region. Look at the map and notice that the city of Mumbai is selected—it is highlighted blue. The graphs display rainfall information for the selected city—Mumbai.

The data in this lesson gives rainfall amounts in millimeters. The following table shows what some measurements in millimeters would be equivalent to in inches (25.4 mm = 1 in.).

mm	100	200	600	1,600	2,800	5,600	12,000
inches	3.9	7.9	23.6	63.0	110.2	220.5	472.4

> Answers to questions in this activity should be recorded on the answer sheet.

(Q1) *Which month gets the most rainfall in Mumbai?*

(Q2) *Which months appear to get little or no rainfall in Mumbai?*

(Q3) *Approximately how much rainfall does Mumbai get each year (in millimeters)?*

(Q4) *Write a sentence summarizing the overall pattern of rainfall in Mumbai in an average year.*

1. At the bottom of the table of contents, click the Selection tab. Click the boxes to uncheck all the layers except Cities.

   ☑ **Cities (1)**
   ☐ Rivers
   ☐ Country Borders
   ☐ Physical Features
   ☐ Countries

2. Click the Display tab to return to the table of contents.

3. Click the Select Features tool. Click a dot for another city. (If you get a graph with multiple colored bars, click directly on the dot again to get data for just that city.)

   (Q5) *How did this change the map?*

   (Q6) *How did this change the graphs?*

4. Click the city of Mangalore to select it in the map.

Q7 *Analyze the graphs and fill in the Mangalore section of the table on your answer sheet. Estimate the rainfall amounts.*

5. Hold down the Shift key and click the cities of Mumbai and Ahmadabad.

    To enlarge the graphs and make them easier to read, drag any border with your mouse.

Q8 *Complete the table on the answer sheet. Use estimates.*

Q9 *As you move northward along the subcontinent's west coast, how does the pattern of rainfall change?*

Q10 *Although the monthly rainfall amounts differ, what similarities do you see among the overall rainfall patterns of these three cities?*

Step 3: Compare coastal and inland cities

1. Make sure the Select Feature tool is still active and select Bangalore.

Q11 *Use the Monthly Rainfall and Annual Rainfall graphs to complete the table on your answer sheet.*

2. Hold down the Shift key and select Mangalore.

Q12 *How does the rainfall pattern of Bangalore compare with that of Mangalore?*

3. Click the Measure tool. The Measure window opens, and your cursor turns into a right-angle ruler with crosshairs.

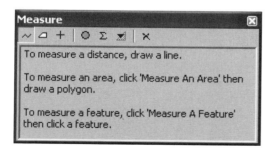

4. Click the Choose Units button, point to Distance, and click Kilometers.

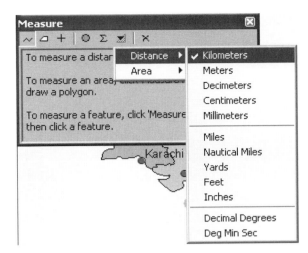

5. Click the dot that represents Bangalore once, and then move the cursor to the dot that represents Mangalore and double-click.

If you accidentally clicked the wrong spot, you can double-click to end the line and start over.

The Measure window reports the distance in kilometers. Segment and length values are the same because only one segment is being measured.

(Q13)   *What is the distance between the two cities?*

6. Close the Measure window.

Although Bangalore is located only a short distance inland from Mangalore, it receives far less rainfall than the coastal city.

7. Turn on the Physical Features layer.

(Q14)   *How can the data in the Physical Features layer help you explain the differences between patterns of rainfall in inland Bangalore and coastal Mangalore?*

8. Turn off Physical Features.

Step 4: Compare eastern and western South Asian cities

1. Click the Select Features tool. One at a time, select the Afghan cities of Kabul and Herat.

   Q15 *Analyze the graphs and complete the table on your answer sheet.*

   Your first impression may be that the Afghan cities get a fair amount of rainfall. But notice that the millimeters scale along the left side of each graph (y-axis) changed to reflect the rainfall range of the selected cities.

   Q16 *Describe the pattern of rainfall in these two cities.*

   Q17 *How do you think Afghanistan's rainfall pattern affects the way of life in that country?*

2. Select the eastern cities of Kolkata and Dhaka.

   Q18 *Analyze the graphs and complete the table on your answer sheet.*

   Q19 *Describe the pattern of rainfall in these two cities.*

3. Hold down the Shift key and select four cities: Herat, Delhi, Kolkata, and Dhaka.

   Q20 *What is happening to the patterns of rainfall as you move from west to east across South Asia?*

4. Click the Clear Selected Features button to unselect the four cities.

5. Close the Monthly Rainfall and Annual Rainfall graphs.

Step 5: Observe yearly precipitation

You've already looked at the monthly precipitation patterns for individual cities across South Asia. In this step, you will add data and look at the total yearly rainfall for regions of South Asia.

1. Click the Add Data button.

2. Navigate to the module 3 LayerFiles folder (**OurWorld2\Mod3\Data\LayerFiles**). Select **Yearly Rain.lyr**.

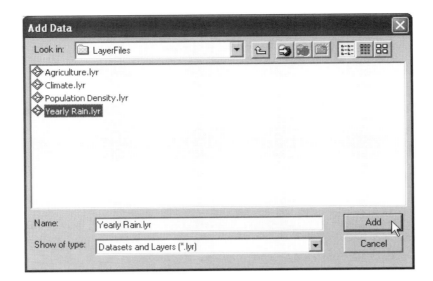

3. Click Add.

4. Drag the layer down in the table of contents so that it is just below the Country Borders layer.

5. Turn off the Cities layer and turn on Yearly Rain (if it's not turned on already).

 Click the Refresh Map button at the bottom of the map area if some city names are still displayed after you turn off the layer.

Q21 *Which regions of South Asia get the least rainfall?*

Q22 *Which regions of South Asia get the most rainfall?*

Q23 *In Q20 you were comparing Herat, Delhi, Kolkata, and Dhaka. Does the map of yearly rainfall that is on your screen now reflect the observation you made at that time? Explain.*

6. Turn off Yearly Rain and turn on Physical Features.

M3
L2

Q24 *What relationships do you see between South Asia's patterns of yearly rainfall and its physical features?*

Step 6: Explore the monsoon's impact on agriculture and population density

The rain patterns and physical features of an area have a significant impact on the way of life of the people who live there. Now you will look at those characteristics and determine the kinds of impact they have on individual countries.

1. Turn on the Country Borders layer.

2. Turn the Physical Features, Rivers, and Yearly Rain layers on and off to make your observations and to answer the questions below.

    Q25 *Which regions or countries of South Asia are suitable for crop farming and which are not? Explain.*

    Q26 *In which regions of South Asia do you expect to see the lowest population density? Explain.*

    Q27 *In which regions of South Asia do you expect to see the highest population density? Explain.*

3. Turn off Physical Features, Rivers, and Yearly Rain layers.

    Now you will add agricultural data for the region and will see if your predictions are correct.

4. Click the Add Data button. Navigate to the module 3 LayerFiles folder (**OurWorld2\ Mod3\Data\LayerFiles**). Select **Agriculture.lyr** and click Add.

5. Drag Agriculture down in the table of contents so that it is just below the Country Borders layer. Turn on the Agriculture layer (if it's not turned on already).

    Q28 *Does the Agriculture layer reflect the predictions you made in Q25? Explain.*

    Q29 *Why are grazing, herding, and oasis agriculture the major activities in Afghanistan?*

    Q30 *What do you know about rice cultivation that would help explain its distribution on the agriculture map?*

    Q31 *Is there any aspect of the agriculture map that surprised you? Explain.*

6. Turn off the Agriculture layer.

You will now examine population density in relation to precipitation and land use.

 7.   Click the Add Data button. Navigate to the module 3 LayerFiles folder (**OurWorld2\ Mod3\Data\LayerFiles**). Add **Population Density.lyr**.

8.   Drag Population Density below Country Borders in the table of contents.

9.   Turn on Population Density (if it's not turned on already).

(Q32) *Does the Population Density layer reflect the population predictions you made in Q26 and Q27? Explain.*

(Q33) *Why is Afghanistan's population density so low?*

(Q34) *Since most of Pakistan gets little to no rainfall, how do you explain the areas of high population density in that country?*

(Q35) *What is the relationship between population density and patterns of precipitation in South Asia?*

(Q36) *What is the relationship between population density and physical features in South Asia?*

10.   Ask your teacher for instructions on where to save this ArcMap map document and on how to rename it.

11.   If you are not going to save the map document, exit ArcMap by choosing Exit from the File menu. When asked if you want to save changes to **Region3.mxd** (or **Region3**), click No.

In this lesson, you explored the patterns of monsoon rainfall in South Asia. You used ArcMap to compare monthly and annual patterns of precipitation in cities throughout the region and explore the relationship between those patterns and the region's physical features. After analyzing this data, you added layers reflecting patterns of agriculture and population density and analyzed the relationship between those human characteristics and the region's climate and landforms.

Name_____   Date_____

Module 3, Lesson 2

Seasonal differences

Step 2: Observe patterns of rainfall

Q1)   Which month gets the most rainfall in Mumbai?

_____

Q2)   Which months appear to get little or no rainfall in Mumbai?

_____

Q3)   Approximately how much rainfall does Mumbai get each year (in millimeters)?

_____

Q4)   Write a sentence summarizing the overall pattern of rainfall in Mumbai in an average year.

_____

_____

_____

Q5)   How did this change the map?

_____

_____

Q6)   How did this change the graphs?

_____

_____

Q7) Analyze the graphs and fill in the Mangalore section of the table below.

City	Months with rainfall > 50mm	Highest monthly rainfall (mm)	Total annual rainfall (mm)
Mangalore			
Mumbai			
Ahmadabad			

Q8) Complete the rest of the table above. Use estimates.

Q9) As you move northward along the subcontinent's west coast, how does the pattern of rainfall change?

_____

_____

Q10) Although the monthly rainfall amounts differ, what similarities do you see among the overall rainfall patterns of these three cities?

_____

_____

Step 3: Compare coastal and inland cities

Q11) Complete the table below.

City	Months with rainfall > 50 mm	Highest monthly rainfall (mm)	Total annual rainfall (mm)
Bangalore			

Q12) How does the rainfall pattern of Bangalore compare with that of Mangalore?

Similarities: _____

_____

Differences: _____

_____

Q13) What is the distance between the two cities? _____

Q14) How can the data in the Physical Features layer help you explain the differences between patterns of rainfall in inland Bangalore and coastal Mangalore?

_____

_____

_____

_____

_____

_____

_____

_____

Step 4: Compare eastern and western South Asian cities

Q15) Analyze the graphs and complete the table below.

City	Months with rainfall ≥ 20 mm	Highest monthly rainfall (mm)	Total annual rainfall (mm)
Kabul			
Herat			

Q16) Describe the pattern of rainfall in these two cities.

_____

_____

Q17) How do you think Afghanistan's rainfall pattern affects the way of life in that country?

_____

_____

Q18) Analyze the graphs and complete the table below.

City	Months with rainfall ≥ 20mm	Highest monthly rainfall (mm)	Total annual rainfall (mm)
Kolkata			
Dhaka			

Q19) Describe the pattern of rainfall in these two cities.

_____

_____

Q20) What is happening to the patterns of rainfall as you move from west to east across South Asia?

_____

_____

Step 5: Observe yearly precipitation

Q21) Which regions of South Asia get the least rainfall?

_____

_____

Q22) Which regions of South Asia get the most rainfall?

_____

_____

Q23) In Q20 you were comparing Herat, Delhi, Kolkata, and Dhaka. Does the map of yearly rainfall that is on your screen now reflect the observation you made at that time? Explain.

_____

_____

_____

_____

Q24) What relationships do you see between South Asia's patterns of yearly rainfall and its physical features?

_____

_____

_____

_____

Step 6: Explore the monsoon's impact on agriculture and population density

Q25) Which regions or countries of South Asia are suitable for crop farming and which are not? Explain.

_____

_____

_____

_____

Q26) In which regions of South Asia do you expect to see the lowest population density? Explain.

_____

_____

_____

_____

Q27) In which regions of South Asia do you expect to see the highest population density? Explain.

_____

_____

_____

_____

Q28) Does the Agriculture layer reflect the predictions you made in Q25? Explain.

_____

_____

_____

_____

Q29) Why are grazing, herding, and oasis agriculture the major activities in Afghanistan?

_____

_____

_____

_____

Q30) What do you know about rice cultivation that would help explain its distribution on the agriculture map?

_____

_____

_____

_____

Q31) Is there any aspect of the agriculture map that surprised you? Explain.

_____

_____

_____

_____

Q32) Does the Population Density layer reflect the population predictions you made in Q26 and Q27? Explain.

_____

_____

_____

_____

Q33) Why is Afghanistan's population density so low?

_____

_____

_____

_____

Q34) Since most of Pakistan gets little to no rainfall, how do you explain the areas of high population density in that country?

_____

_____

_____

_____

Q35) What is the relationship between population density and patterns of precipitation in South Asia?

_____

_____

_____

_____

Q36) What is the relationship between population density and physical features in South Asia?

_____

_____

_____

_____

Middle school assessment
Module 3, Lesson 2
Seasonal differences

Assume the role of an American student who is spending a year living in South Asia as an exchange student. Write four letters to friends or family back home about your experiences and observations during your year in South Asia. Your four letters should be dated January 1, April 1, July 1, and October 1. Using the ArcMap map document as a guide, describe seasonal changes in your city and ways that your daily life and the lives of people around you reflect those changes. You may choose to spend your year in or near one of these cities: Mumbai, Kolkata, or Dhaka. You may use additional sources such as your geography book, encyclopedias, and the Internet to help you develop your letters.

Use the space below to brainstorm for your essay.

M3
L2

January 1
April 1
July 1
October 1

High school assessment
Module 3, Lesson 2
Seasonal differences

Assume the role of an American student who is spending a year traveling in South Asia. Write four letters to friends or family back home about your experiences and observations during your year abroad. Each letter should be written from a different South Asian city. Your four letters should be dated January 1, April 1, July 1, and October 1. Using the ArcMap map document as a guide, describe seasonal characteristics of each city on the date you are writing and ways that your daily life and the lives of people around you reflect those characteristics. You may use additional sources such as your geography book, encyclopedias, and the Internet to help you develop your letters.

Use the space below to brainstorm for your essay.

January 1
April 1
July 1
October 1

Module 4, Lesson 1

The march of time

A global perspective

- Activity
- Answer sheet
- Assessment
- Assessment graph

Module 4, Lesson 1

The march of time

In this activity you will use GIS to identify the world's largest cities at different times during the past 2000 years. You will look for patterns in their locations and speculate on reasons for changes in the patterns.

Step 1: Open a map document

1. Double-click the ArcMap icon on your computer's desktop.

2. When the ArcMap start-up dialog box appears, click **An existing map** and click OK.

M4
L1

3. Navigate to the module 4 folder (**OurWorld2\Mod4**) and choose **Global4.mxd** (or **Global4**) from the list.

4. Click Open. When the map document opens, you see a world map. The table of contents has a data frame called March of Time.

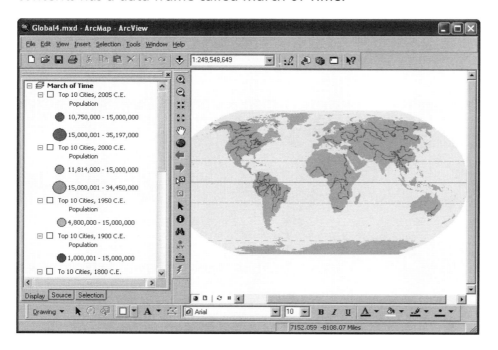

Step 2: Look at cities in 100 CE

1. Scroll down the table of contents on the left side of the ArcMap window until you see the layer called Top 10 Cities, 100 CE (CE stands for Common Era). Click the box to the left of the name to turn on the layer.

> Answers to questions in this activity should be recorded on the answer sheet.

Q1 *Where are the ten largest cities in the world in 100 CE located on the earth's surface?*

Q2 *Where are they located in relation to each other?*

Q3 *Where are they located in relation to physical features?*

Q4 *What are possible explanations for the patterns you see on this map?*

Step 3: Find historic cities and identify modern cities and countries

You will use the Find tool to locate the historic cities on the map.

1. Click the Find tool.

2. Type Carthage in the Find dialog box, and then click Find.

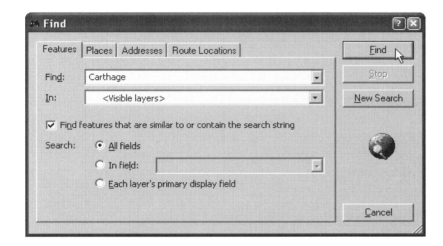

The Find dialog box expands to show a results box. A row for Carthage is listed in the box at the bottom.

3. Move the Find dialog box so you can see all of the cities on your map. Then right-click the listed row for Carthage and click Select. The dot that represents Carthage is highlighted blue on the map.

To complete the table in the answer sheet, you need to identify the modern city located in the same place.

4. Right-click the row for Carthage again, and this time click Identify. The information that you need to complete the table on the answer sheet appears in the Identify window.

M4
L1

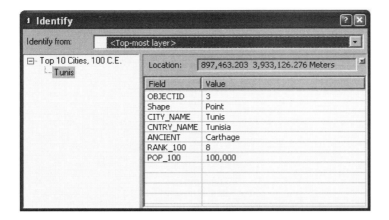

5. Move the Identify window so it is off the Find dialog box and the map. (To move the window, click its title bar and drag it out of the way.)

Q5 *Use the Find and Identify tools to complete the table on the answer sheet.*

6. Close the Identify and Find windows.

7. Click the Clear Selected Features button on the Tools toolbar.

Step 4: Find the largest city of 100 CE and label it

Q6 *What's your estimate of how many people lived in the world's largest city in 100 CE?*

1. Right-click the Top 10 Cities, 100 CE layer in the table of contents, and choose Zoom to Layer. The map zooms to the region of the world where these cities are located.

2. Right-click the Top 10 Cities, 100 CE layer again and choose Open Attribute Table. Each row in this table is associated with one of the city points on the map.

> If you see a lot of extra gray area under the last record or to the right of the table, drag the bottom edge of the window up or drag the edge of the window to the left so that the table takes up less space on your screen.

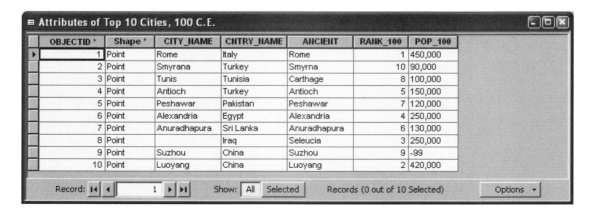

OBJECTID	Shape	CITY_NAME	CNTRY_NAME	ANCIENT	RANK_100	POP_100
1	Point	Rome	Italy	Rome	1	450,000
2	Point	Smyrana	Turkey	Smyrna	10	90,000
3	Point	Tunis	Tunisia	Carthage	8	100,000
4	Point	Antioch	Turkey	Antioch	5	150,000
5	Point	Peshawar	Pakistan	Peshawar	7	120,000
6	Point	Alexandria	Egypt	Alexandria	4	250,000
7	Point	Anuradhapura	Sri Lanka	Anuradhapura	6	130,000
8	Point		Iraq	Seleucia	3	250,000
9	Point	Suzhou	China	Suzhou	9	-99
10	Point	Luoyang	China	Luoyang	2	420,000

Record: 1   Show: All | Selected   Records (0 out of 10 Selected)   Options ▼

3. Scroll right and locate the field name POP_100. Click the field name to highlight the whole column.

4. Right-click the field name. Choose Sort Descending to list the cities from largest to smallest in terms of population.

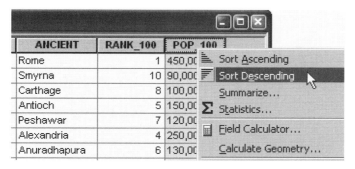

In the POP_100 column, the value –99 for Suzhou indicates that no data is available. It does not mean that there was a population of –99 in Suzhou.

Q7 *What was the largest city in 100 CE ?*

Q8 *What was the population of the world's largest city in 100 CE ?*

5. Click the gray box at the beginning of the row with the largest city. The selected record turns blue in the attribute table, and so does its corresponding dot on the map.

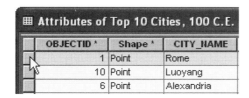

6. Close the Attributes of Top 10 Cities, 100 CE table.

7. Make sure the Draw toolbar is displayed. If it is not, right-click in the gray space next to the Help menu, click Draw, and dock the toolbar at the bottom of the ArcMap window.

8. Click the drop-down arrow to the right of the New Text tool in the Draw toolbar and select the Label tool.

M4
L1

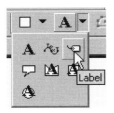

9. Move the Label Tool Options dialog box off the map if necessary. You will accept the default options in the Label Tool Options dialog box.

10. Hover your cursor over the selected city. When the MapTip displays the city name (Rome), click to add the label to the map.

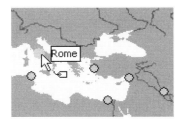

 If you accidentally label the wrong feature, click the Undo button and try again. If you want to reposition the label, use the Select Elements tool to drag the text where you want it to go. If you want to change the font, size, or style of the text, use the options on the Draw toolbar.

11. Click the Select Elements tool and click anywhere on the map away from the label to remove the selection box around the label. The Label Tool Options dialog box also closes.

12. Click the Clear Selected Features button.

13. Click the Full Extent button to see the entire world in the map.

Step 5: Look at cities in 1000 CE and label the most populous city

1. Turn on the Top 10 Cities, 1000 CE layer.

    ☐ ☐ Top 10 Cities, 1500 C.E.
            Population
           ● <1,000,000
    ☐ ☑ Top 10 Cities, 1000 C.E.
            Population
           ○ < 1,000,000
    ☐ ☑ Top 10 Cities, 100 C.E.
            Population
           ○ < 1,000,000

2. Right-click the Top 10 Cities, 1000 CE layer and choose Zoom to Layer.

3. Look for similarities and differences between these points and the cities of 100 CE in the locations and distribution of the world's largest cities.

Q9 *What notable changes can you see from 100 CE to 1000 CE?*

Q10 *What similarities can you see between 100 CE and 1000 CE?*

Now you will look up and label the most populous city of 1000 CE.

4. Right-click the Top10 Cities, 1000 CE layer and choose Open Attribute Table. This shows you all the attribute data associated with the yellow dots on the map.

5. Locate the POP_1000 field. Click the field name to highlight the column.

6. Right-click the POP_1000 name and choose Sort Descending to list the cities from largest to smallest in terms of population.

Q11 *What was the largest city in 1000 CE?*

Q12 *What was the population of the world's largest city in 1000 CE?*

7. Click the gray box at the beginning of the row with the world's largest city in the attribute table. The selected row turns blue in the attribute table, and its corresponding dot on the map turns blue.

8. Close the Attributes of Top 10 Cities, 1000 CE table.

9. Click the Label tool. Click the selected city on the map to label it.

10. Click the Select Elements tool.

11. Reposition the text and use the options on the Draw toolbar to change the font, size, or style of the label text, if desired.

12. Click anywhere in the ocean with the Select Elements tool to unselect the text.

13. Click the Clear Selected Features button to make all the cities in 1000 CE yellow again.

Step 6: Compare other historical periods and formulate a hypothesis

1. Click the Full Extent button.

2. One at a time, turn on each of the remaining six layers representing the years 1500, 1800, 1900, 1950, 2000, and 2005.

Q13 *Complete the table on the answer sheet. Refer back to Q7–Q12 to get information for 100 CE and 1000 CE. (You may need to turn layers on and off or move them up or down in the table of contents.)*

Q14 *Using the map document and your answers in Q13, identify historical periods associated with the greatest changes and provide possible explanations for the changes.*

Step 7: Investigate cities in the present time

As a class, before you began the lesson, you made a guess about the top 10 cities in the world today. You will now see if any of your predictions are correct.

1. Zoom to the full extent of the map and leave only the Top 10 Cities, 2005 CE layer on.

Q15 *How many of your original guesses are among the Top 10 Cities, 2005 CE?*

Q16 *Which cities did you successfully guess?*

Most likely, there are cities on your list that are not in the Top 10 Cities, 2005 CE layer. In order to look at population data for these other cities, you will add a layer with the top 30 cities.

2. Click the Add Data button.

3. Navigate to the module 4 LayerFiles folder (**OurWorld2\Mod4\Data\LayerFiles**). Select **Top 30 Cities 2005.lyr** and click Add.

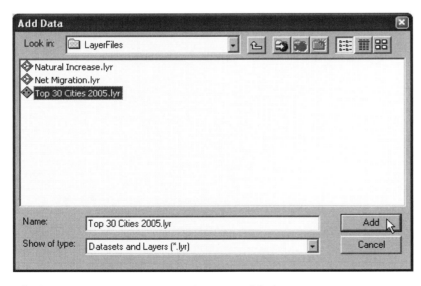

The Top 30 Cities, 2005 CE layer is added to your map.

4. Use the Find tool to locate one of the other cities that you guessed in the beginning of this activity. Make sure to select Top 30 Cities, 2005 CE as the layer to search, as in the example pictured on the following page.

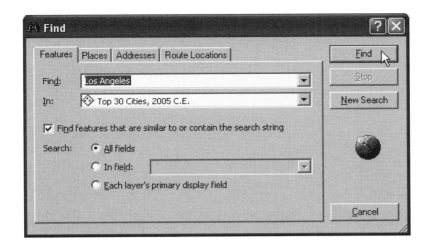

5. Right-click in the results box on the row for the city you found and choose Identify. The Identify dialog box appears with information for the city you found.

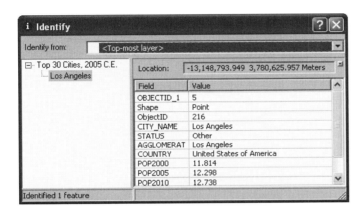

Q17 *In the table on your answer sheet, write the city's name, 2005 population, and rank. You can determine the rank by opening the attribute table and sorting on the POP2005 field. Rank the largest city as 1.*

Q18 *Continue to fill out the table for the other cities in your list or cities that interest you. If you have cities on your list that are not in the top 30, fill in the name, leave the Population column blank, and write >30 in the Rank column.*

6. When you are finished, close the Identify window and the table, if it is open.

Q19 *In general, how far are these other cities from the top 10?*

7. Ask your teacher for instructions on where to save this map document and on how to rename it.

8. If you are not going to save the map document, exit ArcMap by clicking the File menu and clicking Exit. When asked if you want to save changes to **Global4.mxd** (or **Global4**), click No.

In this exercise, you explored population data from 100 CE through the year 2005. You used ArcMap to find, identify, and label the world's most populous cities during different time periods. After analyzing this data, you added data for more large cities and explored populations of your cities of interest.

Name_____   Date_____

Module 4, Lesson 1

The march of time

Step 2: Look at cities in 100 CE

Q1)   Where are the ten largest cities in the world in 100 CE located on the earth's surface?

_____

_____

_____

_____

Q2)   Where are they located in relation to each other?

_____

_____

_____

_____

Q3)   Where are they located in relation to physical features?

_____

_____

_____

_____

Q4) What are possible explanations for the patterns you see on this map?

_____

_____

_____

_____

_____

_____

_____

Step 3: Find historic cities and identify modern cities and countries

Q5) Use the Find and Identify tools to complete the table below.

City		Modern country
Historic name	Modern name	
Carthage		
Antioch		
Peshawar		

Step 4: Find the largest city of 100 CE and label it

Q6) What's your estimate of how many people lived in the world's largest city in 100 CE?

_____

Q7) What was the largest city in 100 CE? _____

Q8) What was the population of the world's largest city in 100 CE? _____

Step 5: Look at cities in 1000 CE and label the most populous city

Q9) What notable changes can you see from 100 CE to 1000 CE?

_____

_____

_____

_____

_____

Q10) What similarities can you see between 100 CE and 1000 CE?

_____

_____

_____

_____

_____

Q11) What was the largest city in 1000 CE? _____

Q12) What was the population of the world's largest city in 1000 CE? _____

Step 6: Compare other historical periods and formulate a hypothesis

Q13)   Complete the table below. Refer back to Q7–Q12 to get information for 100 CE and 1000 CE

Year CE	Largest city	Population of largest city	Major differences in top 10 cities compared with previous time period
100			
1000			
1500			
1800			
1900			
1950			
2000			
2005			

Q14) Using the map document and your answers in Q13, identify historical periods associated with the greatest changes and provide possible explanations for the changes.

Time period of significant change	Explanation for change

Step 7: Investigate cities in the present time

Q15) How many of your original guesses are among the Top 10 Cities, 2005 CE?

_____

Q16) Which cities did you successfully guess?

_____

_____

_____

_____

_____

Q17) In the table on the following page, write the city's name, 2005 population, and rank.

Q18) Continue to fill out the table for the other cities in your list or cities that interest you. If you have cities on your list that are not in the top 30, fill in the name, leave the Population column blank, and write >30 in the Rank column.

City	2005 population	Rank

Q19) In general, how far are these other cities from the top 10?

_____

_____

Middle school assessment
Module 4, Lesson 1

The march of time

A. Use the information you collected in the activity to complete the Assessment graph. Plot a line graph that shows the population of the largest city in each time period. Be sure to label the points and title the graph.

B. Refer to the ArcMap map document (Global4 or the one you saved) and write an essay that compares two of the time periods that you studied. You may use additional sources such as your history and geography books, encyclopedias, and the Internet to help you with your comparisons. On a separate piece of paper, address the following questions in your essay:

1. What was the primary means of transportation in each time period?
2. How did trade between various cities influence the locations of places?
3. How did advancements in technology affect the locations of places?
4. How did distances between major cities change throughout time?
5. What physical features (elevation, proximity to water, etc.) played important roles in the locations of cities?

Use the remainder of this page to brainstorm for your essay.

**M4
L1**

High school assessment
Module 4, Lesson 1

The march of time

A. Use the information you collected in the activity to complete the Assessment graph. Plot a line graph that shows the population of the largest city in each time period. Be sure to label the points and title the graph.

B. Refer to the ArcMap map document (Global4 or the one you saved) and write an essay that compares three of the time periods that you studied. Use at least three additional sources such as your history and geography books, encyclopedias, and the Internet to help you with your comparisons. On a separate piece of paper, address the following questions in your essay:

1. What was the primary means of transportation in each time period?
2. How did trade between various cities influence the locations of places?
3. What physical features (elevation, proximity to water, etc.) played important roles in the locations of cities?
4. Are there any unusual shifts in population centers that you see from one period to the next? What do you think caused these changes?
5. How did advancements in technology affect the locations of places?
6. Predict how future advancements in technology may affect the locations of population centers in the next hundred years.

Use the remainder of the page to brainstorm for your essay.

**M4
L1**

Assessment graph: The march of time

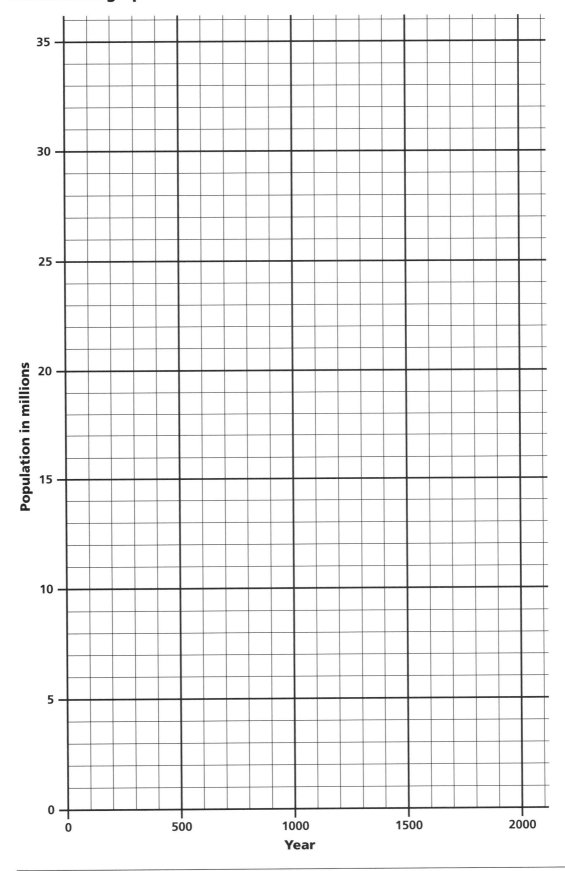

M4
L1

Module 4, Lesson 2

Growing pains

A regional investigation of Europe and Africa

- Activity
- Answer sheet
- Assessment

Module 4, Lesson 2

Growing pains

In this activity, you will analyze natural population growth for different countries. You will focus on Africa, one of the fastest growing regions in the world, and Europe, the slowest growing region in the world. You will analyze the standard-of-living indicators for each region and form a hypothesis about the relationship between these indicators and population growth.

Step 1: Open a map document

1. Double-click the ArcMap icon on your computer's desktop.

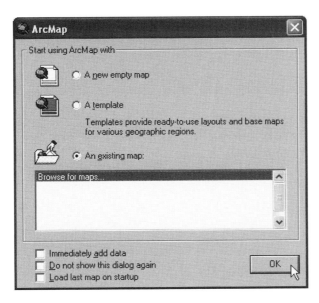

M4
L2

2. When the ArcMap start-up dialog box appears, click **An existing map** and click OK.

3. Navigate to the module 4 folder (**OurWorld2\Mod4**) and choose **Region4.mxd** (or **Region4**) from the list.

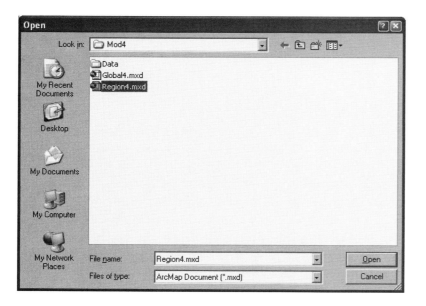

4. Click Open.

5. When the map document opens, click the plus sign next to Population Growth to expand the data frame legend in the table of contents.

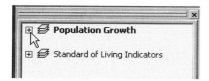

You see a world map with two layers turned on (Countries and Ocean). The check mark next to a layer name tells you the layer is turned on and visible on the map. Two layers, Birth Rate and Death Rate, are listed in the table of contents but are not turned on.

Step 2:   Compare birth rate and death rate data

The world's population is growing because there are more births than deaths each year. This fact can be expressed as a simple formula:
Birth rate – Death rate = Natural increase
(BR – DR = NI)

You will now compare birth rates and death rates around the world to see if you can identify the regions that are growing fastest and slowest.

World Vital Events

World Vital Events Per Time Unit: 2007

(Figures may not add to totals due to rounding)

Time unit	Births	Deaths	Natural increase
Year	133,201,704	55,490,538	77,711,166
Month	11,100,142	4,624,212	6,475,931
Day	364,936	152,029	212,907
Hour	15,206	6,335	8,871
Minute	253	106	148
Second	4.2	1.8	2.5

Source: U.S. Census Bureau, International Data Base.

This table from the U.S. Census Bureau shows the number of births, number of deaths, and rate of natural increase for the world population per year, month, day, hour, and second for 2007.

1. Click the box to the left of the Countries layer in the table of contents to turn the layer off.

2. Click the box next to Birth Rate to turn on the layer. This layer shows the number of births for every one thousand people in a country.

Answers to questions in this activity should be recorded on the answer sheet.

Q1 *Which world region or regions have the highest birth rates?*

Q2 *Which world region or regions have the lowest birth rates?*

3. Turn on the Death Rate layer by clicking the box to the left of its name.

M4
L2

Q3 *Which world region or regions have the highest death rates?*

Q4 *Which world region or regions have the lowest death rates?*

4. Turn the Death Rate layer off and on to compare the two sets of data.

Q5 *If the overall rate of growth is based on the formula BR – DR = NI, which world regions do you think are growing the fastest?*

Q6 *Which world regions do you think are growing the slowest?*

5. Turn off the Death Rate layer.

 6. You can use the Identify tool to learn more about the birth and death rates of specific countries. Click the Identify tool and choose Birth Rate from the Layers list in the Identify window.

7. Move the Identify window so you can see the map (click the window's title bar and drag the window to the desired location).

8. Move your cursor over an African country where the birth rate is very high. Click the country. Your Identify window should look similar to the one below:

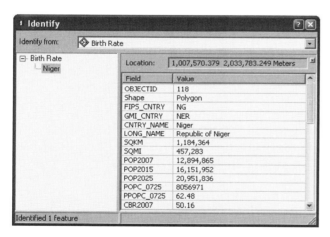

9. Scroll down the list.

This layer contains a lot of information about each country. As you scroll down the list, you see attribute field names in one column and attribute values in another. The birth rate field is abbreviated as CBR2007 (crude birth rate in 2007), and the death rate field is abbreviated as CDR2007 (crude death rate in 2007). In the example above, Niger has a birth rate of 50.16 births per 1,000 living people and a death rate of 20.59 deaths per 1,000 living people.

10. Close the Identify window.

11. Click the Zoom In tool and draw a box around Africa and Europe.

Q7 *Choose two European countries and two African countries and record their birth and death rates in the table on the answer sheet. Use the Identify tool, as described previously, to find information on your chosen countries.*

Q8 *List three questions that the Birth Rate and Death Rate maps raise in your mind.*

12. Close the Identify window.

Step 3: Add the Natural Increase layer

You can test your predictions of the fastest and slowest growing regions (Q5 and Q6) by adding the Natural Increase layer to your map. This layer shows the yearly increase in population that results from the difference between births and deaths in each country.

1. Click the Add Data button.

2. Navigate to the LayerFiles folder within the module 4 Data folder (**OurWorld2\ Mod4\Data\LayerFiles**). Select **Natural Increase.lyr** from the list and click Add.

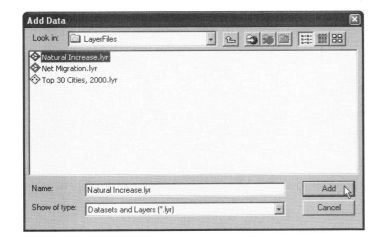

3. Click the Full Extent button to return to the view of the entire world (if your map looks incomplete, click the Refresh View button below the map).

Similarly to the birth and death rates, the natural increase rate is expressed as a specific number of people per 1,000. This means that all the countries colored dark blue on the map add between 24 and 36 people to their populations each year for every 1,000 people already there.

> The actual growth rate of an individual country is based on its natural increase plus the net migration of people into or out of that country each year.

Q9 *What is happening to the populations of countries that are pink?*

Q10 *Which world regions are growing the fastest?*

Q11 *Which world regions are losing people or not growing?*

Q12 *Think about what it would mean for a country to have a population that is growing rapidly or one that is growing slowly or shrinking. Which of these two situations do you think would cause more problems within the country? List some of the problems you would expect to see.*

4. Click the Go Back To Previous Extent button to return the view to Africa and Europe.

Step 4: Look at standard-of-living indicators for Europe and Africa

Geographers look at certain key statistics when they want to compare the standards of living in different countries. They refer to these statistics as "indicators" because they typically reveal or provide some information about the quality of life in that country. The indicators that you will look at in this activity are the following:
- Population 65 years old and over (percent of the total population)
- GDP per capita (annual gross domestic product divided by total population)
- Infant mortality rate (annual number of deaths of infants under one year of age per 1,000 live births)
- Life expectancy (number of years a newborn infant would live if prevailing conditions of mortality at the time of birth continue)
- Literacy rate (percentage of the population over 15 years of age that can read and write; this definition varies slightly between countries)
- Services (percentage of the workforce that is employed in the service sector)

1. Click the minus sign next to Population Growth in the table of contents to collapse the data frame legend.

2. Click the plus sign next to Standard of Living Indicators to expand this data frame legend.

3. Right-click the Standard of Living Indicators data frame name and choose Activate.

4. Click View on the Main menu, point to Bookmarks, and click Europe and Africa.

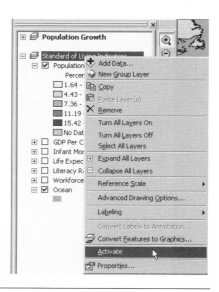

The Standard of Living Indicators map is focused on Europe and Africa. Europe is one of the slowest growing regions and sub-Saharan Africa is one of the fastest growing regions in the world.

5.   Look down the table of contents to see the six standard-of-living indicator layers. The first indicator, Population 65 years or older, shows the percentage of each country's population that is 65 years of age or older.

Q13   *Explore each of the six standard-of-living indicators and complete the table on the answer sheet. Keep in mind the following points:*
  • *A layer will cover the one beneath it when it is turned on. You will need to turn layers on and off to see all the indicators.*
  • *You can change the order of the layers by dragging them to a new position in the table of contents.*
  • *You can expand or collapse the layers in the table of contents to show or hide the legends as you examine different layers.*

Step 5:   Add the Net Migration layer

The net rate of migration is a statistic that indicates the number of people per 1,000 gained or lost each year as a result of migration. A negative number indicates that more people are leaving the country than coming in. A positive number means more people are coming to the country than leaving it.

Q14   *In Q13 you compared standard-of-living indicators for Europe and sub-Saharan Africa. Based on your observations of those indicators, which region would you expect to have a negative net migration? A positive net migration? Explain your answers.*

1.   Click the Add Data button.

2.   Navigate to the LayerFiles folder (**OurWorld2\Mod4\Data**) and add **Net Migration.lyr**.

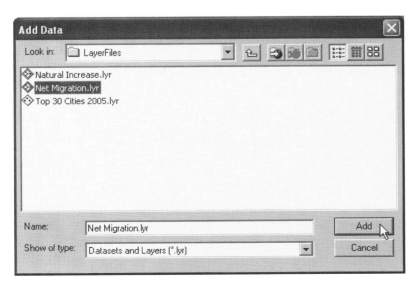

**M4
L2**

Q15 *Summarize the overall patterns of net migration in Europe and sub-Saharan Africa in the table on the answer sheet.*

Q16 *What political or social conditions or events could explain any of the migration patterns you see on the map?*

Step 6: Draw conclusions

1. Click the Layout View button at the bottom of the map area to switch from Data View to Layout View.

    The two data frames are displayed side by side on a layout so you can easily compare the two maps.

2. The Layout toolbar automatically becomes active. If your Layout toolbar is floating, dock it above the map.

3. Expand the Population Growth data frame in the table of contents to display its legend.

4. Enlarge your ArcMap window so that it fills your screen. Click the Zoom Whole Page button on the Layout toolbar.

5. Next, you will be comparing the Natural Increase and Net Migration layers. Make sure that both these layers are visible (not hidden beneath other layers) and that their legends are visible in the table of contents.

    You may want to click the minus sign in front of the other layers to collapse their legends. This way the only visible legends will be the two that you are working with.

6. Click the Zoom In or Zoom Out tool on the Tools toolbar and zoom as needed to focus both maps on Europe and Africa once again (if you accidentally use the Zoom In or Zoom Out tool from the Layout toolbar, click the Zoom Whole Page button).

7. Look at the two maps and compare the rates of natural increase of some countries to their rates of net migration. Think about what correlation, if any, may exist between a country's standard of living, its rate of natural increase, and its rate of net migration.

    Q17 *Based on your map investigations, write a hypothesis about how a country's rate of natural increase affects its standard of living and its net rate of migration.*

Q18 *In the table on the answer sheet, illustrate your hypothesis with data from one European country and one sub-Saharan African country. Use the Identify tool to see the data for an individual country.*

Indicator/Layer	Attribute field
Percentage of population 65 years or older	PPOP_65P
GDP per capita	GDP_PCAP
Infant mortality rate	IMR2007
Life expectancy	LE2007
Literacy rate	LITR_2007
Percent workforce in service sector	PGDP_SV

8. Close the Identify window.

Step 7: Design a layout

You will make some changes to the layout's design before you print it. You will use a template to add a title, legend, and other map elements.

1. Turn off all of the layers in the Population Growth data frame except Natural Increase and Ocean.

2. Click the Change Layout button.

3. Click the General tab in the Select Template dialog box. If your window looks different from the picture below, click the List button in the lower left corner. Select **LetterLandscape.mxt** in the list.

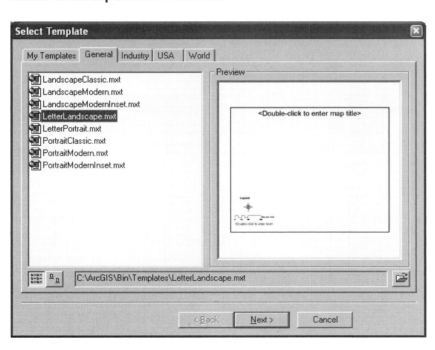

M4
L2

4. Click Next. Make sure the Population Growth data frame is number 1 in the list (if not, click Move Down or Move Up) and click Finish.

    The Population Growth data frame fills the layout. You see text marking the place for a title, a legend, a North arrow, a scale bar, and another place for text.

    > Notice that the Standard of Living Indicators data frame has not been deleted—it has been moved off the layout. You can see part of it in the lower left corner of the layout view.

5. Click the Select Elements tool if it is not already selected. Click the map to display the blue squares at the border of the data frame.

6. Place your cursor over the top middle blue square. When it changes to a double-headed arrow, click and drag the top of the data frame down below the title.

7. Repeat the procedure to drag the left side of the data frame over to the right of the legend and other map elements.

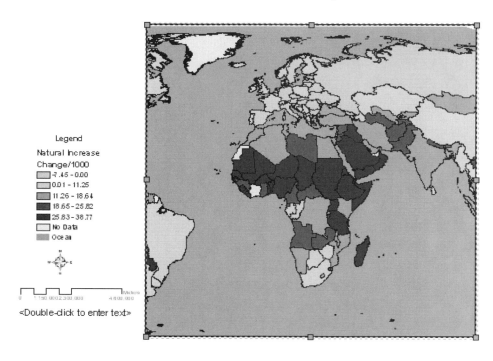

8. Double-click the text that reads <Double-click to enter map title>. Delete the text in the Properties dialog box and type **Population Growth**. Click OK.

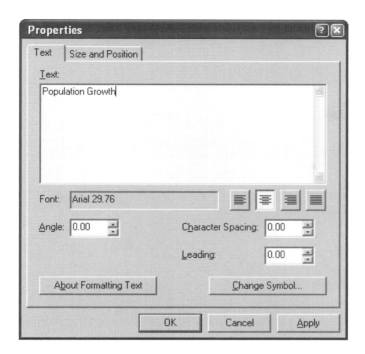

9. Click anywhere on the legend. Click one of the blue corner squares and drag it to make the legend larger.

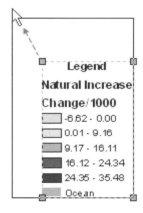

10. Place your cursor in the middle of the legend. Click the legend and hold the mouse button down. Drag the legend up to a suitable location closer to the top of the map.

11. Double-click the compass rose that's below the legend to open the North Arrow Properties dialog box.

12. Click the North Arrow Style button. Choose a different north arrow that you like from the North Arrow Selector. Click OK in both of the dialog boxes to apply your change.

    The new north arrow displays automatically in the layout.

13. Resize the north arrow using the blue squares.

M4
L2

14. Double-click the scale bar to see its properties. Click the Scale and Units tab.

 Q19 *What are the units of measurement?*

By default, ArcMap assigns the units of measurement and intervals based on the coordinate system of the data frame.

15. Click the Division Units drop-down menu and select Miles. Click OK to update the scale bar's units of measurement.

16. Enlarge the scale bar slightly so that it is easier to read.

Step 8: Label your map and print it

When making maps, it's important to include the cartographer's name and the date the map was created. In order to make room for that below the scale bar, you may need to move the north arrow and scale bar up slightly.

1. Adjust the position of the north arrow and scale bar. Click and drag each one to the desired position so that they are centered beneath the legend in the white space to the left of the map.

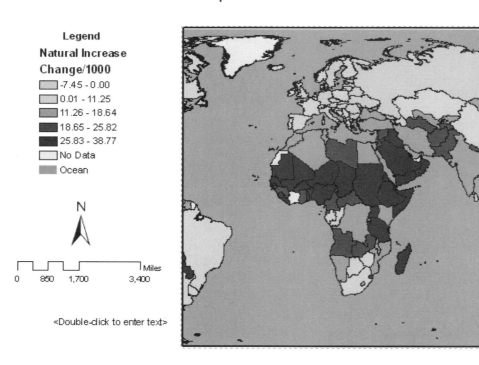

2. To add the name of the cartographer (that's you!) and date, double-click the text below the scale bar.

3. In the text Properties dialog box, type your name. Press Enter to move to the next line and then type today's date. Click OK.

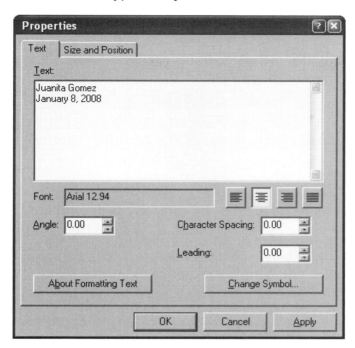

4. Click and drag the text box to the bottom right of the legend area. Then click any blank area on the layout to clear the box around the text element.

Your layout of Population Growth is now ready to print. If you need to make any final adjustments to your map, make them now. Otherwise, proceed to the next step.

M4
L2

5. From the File menu, click Print. Click the Setup button.

6 Use the following settings in the Page and Print Setup dialog box:

- Under Printer Setup, click in the Name box to choose the printer you want to use.
- Under Paper, set the paper size to Letter and the orientation to Landscape.
- Under Map Page Size, check the box to Use Printer Paper Settings.

7. Click OK on both the Page and Print Setup window and the Print window. Your map should print after a few moments.

Step 9: Save your map document

You will be changing the layout. If you want to save your work on this layout, for example to print it or modify it at a later time, you need to save a copy of the map document in its current state.

1. If you wish to save the map document at this point, click the File menu and choose Save As. Ask your teacher for instructions on where to save this map document. Name it ABC_PopGrowth, where ABC are your initials.

2. Choose Save As again and save another copy of the map document for continuing your work on the next layout. Name this copy ABC_StdLiving, where ABC are your intitials.

Step 10: Make a map based on standard-of-living indicators and print it

1. Turn off all layers in the Standard of Living Indicators data frame except Net Migration and Ocean.

2. Click the Change Layout button. Make sure the LetterLandscape.mxt template is selected, and click Next.

3. Click Standard of Living Indicators and then click the Move Up button to move the data frame to the number 1 position. Click Finish. The data frames switch positions in the layout.

4. Follow the procedures you used for your Population Growth map above to now design a Standard of Living presentation map and print it.

5. If you already saved this map document after printing the Population Growth layout, click the Save button to save your work on the second layout. Otherwise, ask your teacher for instructions on where to save this map document and on how to rename it.

6. If you are not going to save the map document, exit ArcMap by choosing Exit from the File menu. When asked if you want to save changes to the map document, click No.

In this lesson, you explored world population growth and analyzed standard-of-living indicators in one of the fastest growing regions of the world (sub-Saharan Africa) and the slowest growing region (Europe). You added layers and worked in two data frames. You created a layout for each data frame and printed your maps.

Name_____   Date_____

Module 4, Lesson 2

Growing pains

Step 2: Compare birth rate and death rate data

Q1) Which world region or regions have the highest birth rates?

_____

Q2) Which world region or regions have the lowest birth rates?

_____

Q3) Which world region or regions have the highest death rates?

_____

Q4) Which world region or regions have the lowest death rates?

_____

Q5) If the overall rate of growth is based on the formula $BR - DR = NI$, which world regions do you think are growing the fastest?

_____

Q6) Which world regions do you think are growing the slowest?

_____

Q7) Choose two European and two African countries and record their birth and death rates in the table below.

Country and continent	Birth rate/1,000	Death rate/1,000
Niger (Africa)	50.16	20.59

Q8) List three questions that the Birth Rate and Death Rate maps raise in your mind.

_____

_____

_____

_____

Step 3: Add the Natural Increase layer

Q9) What is happening to the populations of countries that are pink?

_____

Q10) Which world regions are growing the fastest?

_____

Q11) Which world regions are losing people or not growing?

_____

Q12) Think about what it would mean for a country to have a population that is growing rapidly or one that is growing slowly or shrinking. Which of these two situations do you think would cause more problems within the country?

_____

_____

List some of the problems you would expect to see.

_____

_____

_____

_____

_____

Step 4: Look at standard-of-living indicators for Europe and Africa

Q13) Complete the table below.

Indicator	Compare sub-Saharan Africa and Europe	What does this indicate about the standards of living in these regions?
Population 65 years or older		
GDP per capita		
Infant mortality rate		
Life expectancy		
Literacy rate		
Percent of workforce in service sector		

Step 5: Add the Net Migration layer

Q14) In Q13 you compared standard-of-living indicators for Europe and sub-Saharan Africa. Based on your observations of those indicators, which region would you expect to have a negative net migration?

_____

A positive net migration? _____

Explain your answers. _____

_____

_____

_____

_____

Q15) Summarize the overall patterns of net migration in Europe and sub-Saharan Africa in the table below.

Net migration in sub-Saharan Africa	Net migration in Europe

Q16) What political or social conditions or events could explain any of the migration patterns you see on the map?

_____

_____

_____

_____

_____

Step 6: Draw conclusions

Q17) Based on your map investigations, write a hypothesis about how a country's rate of natural increase affects its standard of living and its net rate of migration.

_____

_____

_____

_____

_____

Q18) In the table below, illustrate your hypothesis with data from one European country and one sub-Saharan African country.

Europe	Data	Africa
	Country name	
	Natural increase	
	Net migration	

Step 7: Design a layout

Q19) What are the units of measurement?

_____

Middle school assessment
Module 4, Lesson 2

Growing pains

You have been selected by the United Nations to establish a model partnership between nations. Your job is to select two countries—one country with a slow growth rate from Europe and one with a fast growth rate from Africa. Refer to the activity if you need help in identifying the growth rate of a country. Select the countries from the list below:

Belgium

Germany

Mozambique*

Somalia*

United Kingdom

Italy

Nigeria

Madagascar*

Zambia

Denmark

Norway

Poland

Tanzania

Write a report that addresses the following points:
1. Identify issues critical to each country in regard to growth and standard of living.
2. How can these countries work together to address their respective problems?
3. Do current relationships (such as trade agreements) exist between these countries, and are these positive or negative? How can these current partnerships be improved?

Support this report with maps and data from the activity. You may also want to refer to additional sources such as textbooks, encyclopedias, and the Internet to find additional details about your countries.

Bonus

Net Migration Rate (NMR) data is unavailable for the three African countries marked with an asterisk (*). What do you guess is the rate of net migration in these countries? Analyze the population growth and standard-of-living indicators in the Region4 map document to help you answer this question.

High school assessment
Module 4, Lesson 2

Growing pains

You have been selected by the United Nations to establish a model partnership among nations. Your job is to select two groups of countries—one group with a slow growth rate and one group with a fast growth rate. Each group of countries should consist of four to five nations from Europe or sub-Saharan Africa. The countries within each group need to have similar growth characteristics. Refer to the activity if you need help identifying the growth rate of a country.

Write a report that addresses the following points:
1. Identify issues related to growth and standard of living that are critical to each group's welfare.
2. How can these countries work together to help improve their standards of living and address their respective problems?
3. Do current relationships (such as trade agreements) exist between these countries, and are these positive or negative? How can these current partnerships be improved?

Support this report with maps and data from the activity. You may also want to refer to additional sources such as textbooks, encyclopedias, and the Internet to find additional details about your countries.

**M4
L2**

Module 5, Lesson 1

Crossing the line

A global perspective

- Activity
- Answer sheet
- Assessment

Module 5, Lesson 1

Crossing the line

Boundaries are invisible lines on the earth's surface. They divide the surface area into distinct political entities. In this activity, you will use GIS to investigate different types of international boundaries, explore the implications of various boundary configurations, and observe boundary changes that occurred in recent years.

Step 1:   Open a map document

1.   Double-click the ArcMap icon on your computer's desktop.

2.   When the ArcMap start-up dialog box appears, click **An existing map** and click OK.

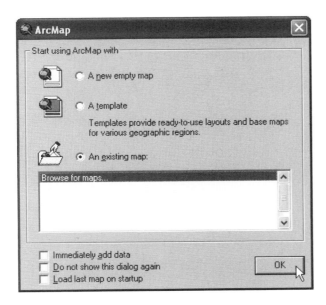

M5
L1

3. Navigate to the module 5 folder (**\OurWorld2\Mod5**) and choose **Global5.mxd** (or **Global5**) from the list.

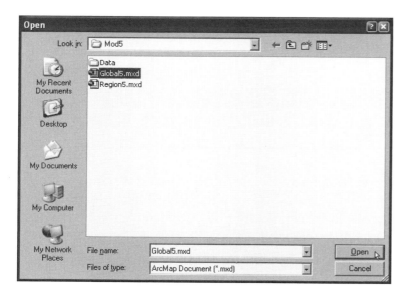

4. Click Open.

The map document opens, and you see a composite satellite image of the world. The check mark next to the layer name tells you the layer is turned on and visible on the map.

Step 2: Explore mountain ranges as physiographic boundaries

As astronaut Russell L. Schweickart said, if you could view the world from space, you would see no boundary lines. Boundaries are human-made lines that define the world's political entities.

There are several types of boundaries between countries. For example, physiographic boundaries follow landscape features such as mountain ranges or rivers.

1. Click the box next to the Boundaries 2007 layer to turn it on. A check mark appears, and the red lines show the international boundaries for the year 2007.

2. Click the Zoom In tool. On the map, click and drag a box around Europe. The view is now centered on Europe.

3. Drag another box around Europe to zoom in more so you can see the physical features in greater detail.

4. Turn off Boundaries 2007 by clicking the check mark next to the layer name.

Locate Europe's mountain ranges in the satellite image. Notice that ranges such as the Pyrenees Mountains in northeastern Spain form a natural boundary. You will use the Draw Line tool to draw a line where you see a mountain range forming a natural boundary between different parts of the continent. First, you will select a symbol type and color for drawing.

5. On the Draw toolbar click the drop-down arrow to the right of the New Rectangle tool and select the New Line tool.

> If the Draw toolbar is not visible, click the View menu and choose Toolbars, Draw to turn it on. A good place to dock it is at the bottom of the ArcMap window.

 6. On the Draw toolbar click the drop-down arrow to the right of the Line Color button and change the color to yellow.

Now you are ready to draw a physiographic boundary in Europe.

7. Click the westernmost edge of the Pyrenees Mountains to start your line. Continue clicking along the path of the mountain range until you reach its easternmost edge (you will see only a faint gray line as you are clicking). Double-click to end the line.

8. The yellow line is displayed in the map. If you don't like your line, you can press the Delete key on your keyboard to delete the line and draw another. When you're finished, click on the map away from the yellow line to make the blue selection box disappear.

9. Turn on the Boundaries 2007 layer. You see that your line corresponds to a border between two countries.

10. Click the Identify button. Move the Identify window so you can see the map.

11. Click the country that borders the Pyrenees Mountains to the north.

M5
L1

The left side of the window shows the layer name and below it the country name. You can also see all the attributes for that country that are in the attribute table.

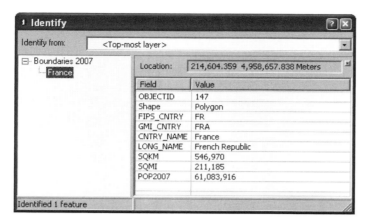

12. Click on the country to the south of the border.

Answers to questions in this activity should be recorded on the answer sheet.

Q1 *The Pyrenees Mountains are the border between which two countries?*

13. Click the Select Elements tool. Click and drag a box over the boundary line you previously drew on the map to select the line (the blue box should reappear). Press the Delete key on your keyboard to delete the line.

14. Use the Identify tool to find other Western European countries where physiographic boundaries created by mountains correspond to actual political boundaries. Click a country to see its information in the Identify window.

Q2 *Complete the table on the answer sheet. Consult an atlas to learn the names of mountain ranges you don't know.*

Step 3: Explore bodies of water as physiographic boundaries

1. Turn off the Satellite Image layer and turn on the Rivers, Lakes, Countries 2007, and Ocean layers.

   Wherever countries have physiographic boundaries based on rivers, the red boundary line disappears beneath the blue river on the map. Look closely at Europe to see if you can find any boundaries that are rivers.

2. In the Identify window click the "Identify from" drop-down menu and make Countries 2007 the reference layer (if the Identify window is closed, click the Identify button to open it).

3. Click a country that has a river as part of a boundary (you may want to zoom or pan the map). Click the Identify tool again when you have found such a country.

Q3 *In the table on the answer sheet, record the names of three pairs of countries that share a boundary that's a river (to identify the names of the rivers, make Rivers the reference layer in the Identify window).*

Coastlines are also physiographic boundaries. Countries that do not have a coastline are said to be landlocked.

4. Zoom or pan so that Western Europe is in your full view.

Q4 *Name three landlocked countries in Western Europe. Use the Identify tool if you don't know the name of a specific country (remember to set Countries 2007 as the reference layer in the Identify window).*

Step 4: Explore geometric boundaries

Another type of boundary is a geometric boundary. Geometric boundaries consist of straight lines that do not correspond to physical features on the earth's surface.

1. Click the Full Extent button to see all the continents.

2. Click the Zoom In tool. Use it to zoom in on Africa.

You see many rivers that overlap boundaries throughout the African continent.

3. Look at the map and locate countries with a shared geometric boundary.

4. Make Countries 2007 the reference layer in the Identify window.

Q5 *List three pairs of countries with a shared geometric boundary.*

5. Close the Identify window.

6. Click Full Extent to see the entire world.

7. Ask your teacher if you should stop here and save this map document. Follow your teacher's instructions on how to rename the map document and where to save it. If you do not need to save the map document, proceed to the next step.

Q6 *Write the new name you gave the map document and where you saved it.*

M5 L1

Step 5: Explore anthropographic boundaries based on language and religion

A third type of boundary is anthropographic. Anthropographic boundaries are based on societal characteristics such as language, religion, or ethnicity.

1. Turn off Rivers, Lakes, Boundaries 2007, and Countries 2007.

2. Click the Add Data button.

3. Navigate to the LayerFiles folder within the module 5 Data folder (**OurWorld2\ Mod5\Data\LayerFiles**).

4. Select **Language.lyr.** Hold down the Ctrl key and click **Religion.lyr.** Click Add.

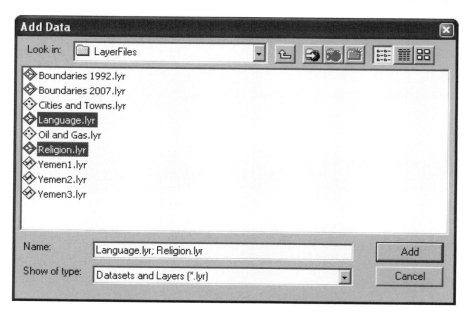

5. Drag the Boundaries 2007 layer above the Language and Religion layers in the table of contents.

6. Turn on Language and click the plus sign next to it to expand the legend. The distribution of major language groups in the world is displayed. Drag the right edge of the table of contents to widen it if you need to.

7. Observe the pattern of anthropographic boundaries based on language.

(Q7) *Use the Identify tool to determine the principal language group in South America and Western Europe (don't forget to make Language the reference layer in the Identify window).*

8. Turn on Boundaries 2007.

(Q8) *Use the Identify and Zoom tools to locate countries separated by an anthropographic boundary based on language. List three pairs of such countries on the answer sheet.*

9. Click Full Extent. Turn off Boundaries 2007 and Language. Click on the minus sign next to Language to collapse the legend.

10. Turn on the Religion layer and expand its legend. The distribution of major religions is displayed.

11. Observe the pattern of anthropographic boundaries based on religion throughout the world.

(Q9) *Use the Identify and Zoom tools to determine the principal religions in North America and Africa. Record them on the answer sheet.*

12. Turn on Boundaries 2007.

(Q10) *Use the Identify and Zoom tools to locate countries separated by an anthropographic boundary based on religion. List three pairs of such countries on the answer sheet.*

Step 6: Review physiographic, geometric, and anthropographic boundaries

(Q11) *List additional examples of countries separated by physiographic, geometric, or anthropographic boundaries for each continent in the table on the answer sheet.*

Step 7: Explore the effects of boundary shape, cultural diversity, and access to natural resources

Boundaries define a country's size and shape, or territorial morphology, which may be related to the cohesiveness of that country. Small compact nations or ones that are circular or hexagonal, for example, are likely to be more united than ones that are elongated or fragmented.

1. Close the Identify window if it's still open.

2. Click the Full Extent button to see the whole world again. Turn off all layers except for Countries 2007 and Ocean and collapse the Religion legend in the table of contents.

3. Click the Find button.

4. In the Find dialog box type **Chile**. Select Countries 2007 from the "In" drop-down menu.

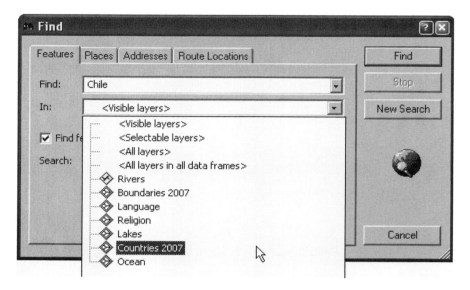

5. Click Find. A results box appears at the bottom of the Find dialog box, and Chile is listed. Move the Find dialog box to the side so you can see the map.

6. Right-click on the highlighted row for Chile and click Zoom To. Then right-click again and click Flash to locate Chile in the map.

(Q12) *The table on the answer sheet illustrates six types of countries based on shape and gives an example of each. Chile is listed as an example of an elongated country. Use the Find, Zoom, and Pan tools to locate another example of each type of country. Record the countries in the Example 2 column. Remember, you can use the Identify tool to find the names of countries that you do not know.*

Another factor that influences cohesiveness is cultural diversity.

7. Click the Full Extent button. Turn off the Countries 2007 layer. Turn on Boundaries 2007 and Language.

(Q13) *Identify three culturally uniform countries on the basis of language group.*

(Q14) *Identify three culturally diverse countries on the basis of language groups.*

Boundaries also influence economic activities. Earlier in this activity, you identified landlocked countries in Western Europe. Historically, these countries were limited in their ability to trade directly with other nations because imports and exports had to pass through other countries en route to their destination.

8. Click the Full Extent button. Turn off Boundaries 2007 and Language. Turn on Countries 2007 and Lakes.

(Q15) *Record an example of a landlocked country for each continent listed in the table on the answer sheet. For a continent that does not have a landlocked country, write "None."*

9. Close the Find dialog box if it is still open.

Boundaries also influence a country's economic activities dependent on access to natural resources.

10. Click the Full Extent button.

11. Click the Add Data button. Navigate to the LayerFiles folder within the module 5 data folder (**OurWorld2\Mod5\Data\LayerFiles**). Double-click **Oil and Gas.lyr**. The locations of oil and natural gas sources around the world are displayed on the map.

12. Use the Zoom In tool to focus on Southeast Asia.

(Q16) *Name two Southeast Asian countries that do not have any oil and gas resources within their land borders (use the Pan, Zoom, and Identify tools).*

(Q17) *Name two Southeast Asian countries that have oil and gas resources within their land borders.*

M5 L1

13. Turn off the Oil and Gas layer.

14. Click the Full Extent button. Turn off Countries 2007 and Ocean, and turn on Boundaries 2007 and Satellite Image.

Step 8: Explore boundary changes that occurred in the 1990s

Political boundaries can change in many ways. A large country may split into smaller ones, small countries may combine to form a larger one, and territories that were once part of one country may be incorporated into another.

1. Click the Add Data button. Navigate to the LayerFiles folder within the module 5 data folder (**OurWorld2\Mod5\Data\LayerFiles**) and add **Boundaries 1992.lyr**.

2. The international boundaries from 1992 are displayed as yellow lines.

Because the 1992 boundaries cover the 2007 boundaries on the map, the only 2007 boundary lines (red) that are visible are those that did not exist in 1992.

Q18 *Describe three political-boundary changes between 1992 and 2007 (use the Zoom and Pan tools to get a closer look, and the Identify tool to identify countries).*

Q19 *Name two countries that existed in 1992 but do not exist in 2007.*

3. If you already saved this map document earlier, click the Save button to save your work. Otherwise, ask your teacher for instructions on where to save this map document and how to rename it. If you do not need to save the map document, continue to the next step.

Q20 *Write the new name you gave the map document and where you saved it.*

Countries in groups A and B below are new countries that have emerged since 1992.

Group A	Group B
Czech Republic	Russia
Slovakia	Belarus
Slovenia	Ukraine
Croatia	Moldova
Bosnia and Herzegovina	Armenia
Serbia	Azerbaijan
Montenegro	Georgia
Macedonia	Kazakhstan
Eritrea	Uzbekistan
	Tajikistan
	Turkmenistan
	Kyrgyzstan

Q21 *Select three countries from group A and three from group B and complete the table on the answer sheet. Use the map and your GIS skills to obtain the information.*

4. Click the File menu and click Exit. When asked if you want to save changes to the map document, click No.

In this exercise, you used ArcGIS to explore the various types of political boundaries and their impact on the countries they define. You added layers and used the Find, Identify, Zoom, and Pan tools to investigate the maps. You observed and analyzed boundary changes between 1992 and 2007.

Name_____ Date_____

Module 5, Lesson 1

Crossing the line

Step 2: Explore mountain ranges as physiographic boundaries

Q1) The Pyrenees Mountains are the border between which two countries?

_____ and _____

Q2) Complete the table below:

Countries that have mountain ranges as political boundaries	Mountains that form the boundary
and	
and	
and	

Step 3: Explore bodies of water as physiographic boundaries

Q3) Record the names of three pairs of countries that share a boundary that's a river.

Countries that have a river as a boundary	River that forms the boundary
and	
and	
and	

Q4) Name three landlocked countries in Western Europe.

_____

_____

_____

Step 4: Explore geometric boundaries

Q5) List three pairs of countries with a shared geometric boundary.

_____ and _____

_____ and _____

_____ and _____

Q6) Write the new name you gave the map document and where you saved it.

Document _____
(Example: ABC_Global5.mxd)

Location _____
(Example: C:\Student\ABC)

Step 5: Explore anthropographic boundaries based on language and religion

Q7) Determine the principal language group in South America and Western Europe.

South America: _____

Western Europe: _____

Q8) Use the Identify and Zoom tools to locate countries separated by an anthropographic boundary based on language. List three pairs of such countries.

_____ and _____

_____ and _____

_____ and _____

Q9) Determine the principal religions in North America and Africa.

North America: _____

Africa: _____

Q10) Use the Identify and Zoom tools to locate countries separated by an anthropographic boundary based on religion. List three pairs of such countries.

_____ and _____

_____ and _____

_____ and _____

Step 6: Review physiographic, geometric, and anthropographic boundaries

Q11) List additional examples of countries separated by physiographic, geometric, or anthropographic boundaries for each continent in the table.

Continent	Countries separated by:		
	Physiographic boundary	**Geometric boundary**	**Anthropographic boundary**
North and Central America	_____ and _____ *Circle one:* Mountains Rivers Lakes	_____ and _____	_____ and _____
South America and the Caribbean	_____ and _____ *Circle one:* Mountains Rivers Lakes	_____ and _____	_____ and _____
Europe	_____ and _____ *Circle one:* Mountains Rivers Lakes	_____ and _____	_____ and _____
Africa	_____ and _____ *Circle one:* Mountains Rivers Lakes	_____ and _____	_____ and _____
Asia	_____ and _____ *Circle one:* Mountains Rivers Lakes	_____ and _____	_____ and _____

Step 7: Explore the effects of boundary shape, cultural diversity, and access to natural resources

Q12) Locate another example of each type of country. Record the countries in the Example 2 column.

Type of country	Example 1	Example 2
Elongated	Chile	
Fragmented	Philippines	
Circular/hexagonal	France	
Small/compact	Bulgaria	
Perforated (has a "doughnut hole")	South Africa	
Prorupted (has a "panhandle")	Namibia	

Q13) Identify three culturally uniform countries on the basis of language group.

_____

_____

_____

Q14) Identify three culturally diverse countries on the basis of language groups.

_____

_____

_____

Q15) Record an example of a landlocked country for each of the following continents. For a continent that does not have a landlocked country, write "None."

Continent	Landlocked country
North and Central America	
South America	
Africa	
Asia	

Q16) Name two Southeast Asian countries that do not have any oil and gas resources within their land borders.

_____

_____

Q17) Name two Southeast Asian countries that have oil and gas resources within their land borders.

_____

_____

Step 8: Explore boundary changes that occurred in the 1990s

Q18) Describe three political-boundary changes between 1992 and 2007.

_____

_____

_____

Q19) Name two countries that existed in 1992 but do not exist in 2007.

_____

_____

Q20) Write the new name you gave the map document and where you saved it.

Document _____
(Example: ABC_Global5.mxd)

Location _____
(Example: C:\Student\ABC)

Q21) Select three countries from group A and three from group B and complete the following table.

Group	Country	Type of boundaries	Shape	Economic advantages or disadvantages	Likelihood of cohesiveness or of splitting apart
A					
B					

Middle school assessment
Module 5, Lesson 1

Crossing the line

Can you predict the future? Instead of using a crystal ball, you will use GIS to see 25 years into the future of the world.

1. Use the information you learned in this lesson to identify a current international boundary that you think will change in the next 25 years.
2. Draw a map (in ArcMap or on paper) to illustrate what this boundary will look like in 25 years. Be sure to include a legend, north arrow, scale, and date of creation on the map.
3. Write an essay that describes the consequences of the change you predict. Address the following questions in your essay:
   - What type of boundary is involved in the projected change?
   - How will the territorial morphologies of the countries involved be affected by the projected change?
   - What will the economic impact of the projected change be?
   - How will the projected change affect internal cohesiveness in the countries involved?

**M5
L1**

High school assessment
Module 5, Lesson 1

Crossing the line

Can you predict the future? Instead of using a crystal ball, you will use GIS to see 25 years into the future of the world.

1. Use the information you learned in this lesson to identify two current international boundaries that you think will change in the next 25 years. One prediction should involve splitting a current country into two or more smaller countries, and the other should merge two or more countries into one larger one.
2. Draw a map (in ArcMap or on paper) to illustrate what these boundaries will look like in 25 years. Be sure to include a legend, north arrow, scale, and date of creation on the map.
3. Write an essay that compares potential consequences of the changes you predict. Address the following questions in your essay:
   - What types of boundaries are involved in the changes?
   - How will the territorial morphologies of the countries involved be affected by the projected changes?
   - What will the economic impact of the projected changes be?
   - How will the projected changes affect internal cohesiveness in the countries involved?

M5
L1

Module 5, Lesson 2

A line in the sand

A regional investigation of Saudi Arabia and Yemen

- Activity
- Answer sheet
- Assessment

Module 5, Lesson 2

A line in the sand

The ever-changing map of the world reflects the forces of conflict and cooperation among nations and peoples of the world. In this activity, you will explore one of the first boundary changes of the twenty-first century—the creation of a new border between Yemen and Saudi Arabia. After more than 60 years of conflict, the two nations signed the historic boundary agreement in June 2000. Using data from the Treaty of Jeddah, you will create a map reflecting the treaty's territory and analyze underlying physiographic and cultural considerations that influenced the location of the boundary.

Step 1: Open a map document

1. Double-click the ArcMap icon on your computer's desktop.

2. When the ArcMap start-up dialog box appears, click **An existing map** and click OK.

3. Navigate to the module 5 folder (**OurWorld2\Mod5**) and choose **Region5.mxd** (or **Region5**) from the list.

**M5
L2**

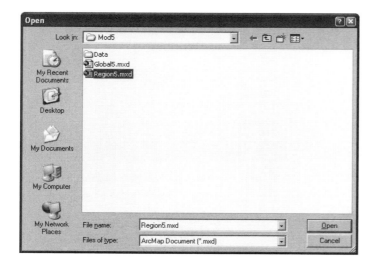

4. Click Open.

    When the map document opens, you see a shaded relief map with the Arabian Peninsula outlined in red.

5. Stretch your ArcMap window so that it fills most of your screen.

Step 2: Identify countries that border the Arabian Peninsula

1. Click the View menu and choose Bookmarks, Arabian Peninsula. Now the Arabian Peninsula fills the view.

2. Look in the table of contents for a layer called "Neighbors - outline." Click the box to the left of the layer name to turn it on.

3. Slide the mouse pointer over the map to display the country names.

> Answers to questions in this activity should be recorded on the answer sheet.

Q1 *What countries border the Arabian Peninsula to the north?*

Step 3: Investigate the physical characteristics of the Arabian Peninsula

The map on your screen is a shaded relief map. It depicts landforms such as mountain ranges, valleys, plateaus, and plains.

Q2 *Is any part of the Arabian Peninsula mountainous?*

Q3 *If so, where are the mountains located?*

1. Click the plus sign next to Water in the table of contents to expand this group of layers.

2. Turn on the Bodies of Water and Streams layers. Then display them by checking the box next to Water.

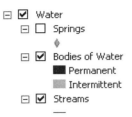

Most of the streams you see on your map are intermittent, which means that they are dry during some parts of the year.

Q4 *Are there any parts of the Arabian Peninsula that do not have any water at all? If so, where are these regions?*

Q5 *Do you see any relationship between landforms and the availability of water?*

3. Turn off the Streams layer and observe the distribution of permanent bodies of water on the Arabian Peninsula.

4. Click the Zoom In tool. Click and drag a small box around an area of blue dots. Now you can see the bodies of water more closely.

Q6 *Describe the bodies of water in terms of their connectedness or disconnectedness.*

5. Click the Previous Extent button to return to your view of the entire peninsula.

6. Turn off the following layers: Water, Arabian Peninsula—outline, Neighbors—outline, and Shaded Relief (you may need to scroll down in the table of contents).

Your map display should look like this:

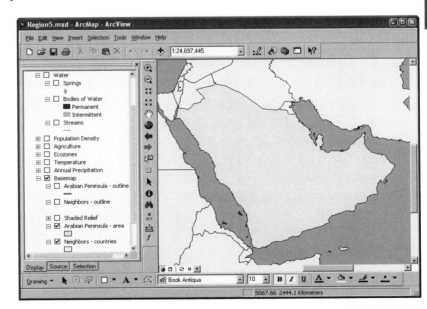

M5
L2

7. Collapse the Water group. Expand the Annual Precipitation layer and turn it on.

Amounts of rainfall are given in millimeters in the map document. The following table shows what some measurements in millimeters would be equivalent to in inches (25.4 mm = 1 in.).

mm	100	200	300	400	500	600	700
inches	3.9	7.9	11.8	15.7	19.7	23.6	27.6

Q7 *A desert is defined as a place that gets less than 10 inches of rain per year. How many millimeters equal 10 inches?*

Q8 *Based on the amounts of rainfall displayed on the map, do you think there is much farming on the Arabian Peninsula? Explain.*

Q9 *Approximately what percentage of the Arabian Peninsula is desert?*

8. Turn off Annual Precipitation. Turn on and expand the Temperature layer group. Turn on Temp: Sept. – Nov.

Temperatures are given in degrees Celsius in the map document. The following table shows what some measurements in degrees Celsius would be equivalent to in degrees Fahrenheit.

°C	5	10	15	20	25	30	35	40
°F	41	50	59	68	77	86	95	104

Q10 *What is the approximate range of temperatures across the Arabian Peninsula during September through November?*

The three layers below "Temp: Sept.- Nov." display temperature information for the periods December–February, March–May, and June–August.

9. Turn the temperature layers on and off one at a time to see the change of temperatures on the Arabian Peninsula through the four seasons.

Q11 *Which season is the hottest?*

Q12 *What is the approximate range of temperatures across the Arabian Peninsula during the hottest season?*

10. Turn off the Temperature group. Turn on the Ecozones layer and expand its legend.

Q13) *What relationship do you see between the Arabian Peninsula's ecozones and its patterns of landforms, precipitation, and temperature?*

Q14) *Use your answers from previous questions and turn different layers on and off to come up with three observations for each physical characteristic in the table on the answer sheet.*

Q15) *In your opinion, which of the region's physical characteristics would be considered "valuable" in a boundary decision? Explain.*

Step 4: Investigate the population characteristics of the Arabian Peninsula

The population of the Arabian Peninsula is approximately 57 million. The majority of this population lives in Saudi Arabia (27.5 million) and Yemen (22 million). The remaining 7.5 million live in Oman, the United Arab Emirates, and Qatar.

1. Turn off Ecozones. Turn on the "Arabian Peninsula – names" layer. This layer identifies the countries by name but does not show their borders (you will explore the borders of these countries later in the activity).

2. Turn on the Major Cities and Agriculture layers. Expand the Agriculture legend to see the types of agricultural activity on the Arabian Peninsula.

Q16) *What is the principal agricultural activity on the peninsula?*

Q17) *Based on what you now know about the physical characteristics of the region, why do you think the agricultural activity is so limited?*

3. Turn on and expand the Population Density layer. The population is displayed as number of people per square kilometer.

Q18 *How does Yemen compare to the rest of the Arabian Peninsula in population density?*

Q19 *Describe the overall population density of the Arabian Peninsula.*

4. Turn off Population Density. Make sure Agriculture is still turned on.

5. Turn on the Water layer, expand it, and turn on Springs. Turn off Bodies of Water. The Springs layer shows the locations of springs and water holes.

```
☐ ☑  Water
     ☐ ☑  Springs
          ◆
     ☐ ☐  Bodies of Water
          ■ Permanent
          ▨ Intermittent
     ☐ ☐  Streams
          —
```

Q20 *Speculate about the ways water is most commonly used at these springs and water holes.*

Q21 *Use your answers from Q16–Q20 and analysis of the maps to list two observations for each population characteristic in the table on the answer sheet.*

Q22 *If an international boundary were to be drawn across some part of the Arabian Peninsula, how would these population characteristics influence the perception of certain regions as being more valuable than others?*

Step 5:   Investigate the Empty Quarter

1. Turn off Springs and turn on Roads. Expand the Roads legend.

   Take note of the large area with practically no roads in the south-central part of the peninsula. This region is called the Rub´ al-Khali and is also known as the Empty Quarter. Saudi Arabian borders with its southern neighbors cross this region.

2. Turn the following layers on and off so you can observe the characteristics of the Empty Quarter: Streams, Population Density, Agriculture, Ecozones, Temperature, and Annual Precipitation.

Q23 *List two observations on the physical characteristics of the Empty Quarter and two observations on its population characteristics.*

Q24 *What difficulties would an area like this present if an international boundary must cross it?*

3. Right-click the Arabian Peninsula data frame in the table of contents and choose Collapse All Layers.

4. Right-click Arabian Peninsula again and choose Turn All Layers Off. Then turn the following layers back on: Major Cities, Arabian Peninsula - names, and Basemap.

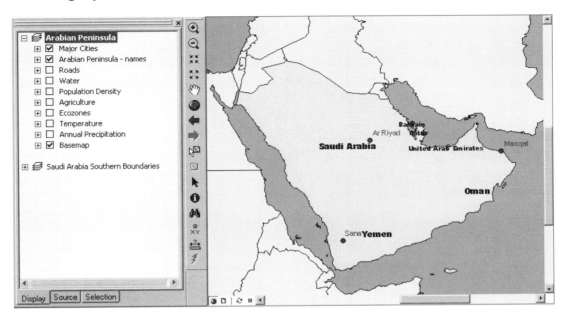

5. Ask your teacher for instructions on where to save this map document and what name to give it. Click the File menu and Save As to navigate to the desired folder and rename the map document.

(Q25) *Write the map document's new name and location.*

6. If you will be stopping here and continuing the lesson at a later time, click the File menu and Exit.

Step 6: Explore Saudi Arabia's southern boundaries

1. If you exited at the end of the last section, start ArcMap, navigate to the folder where you renamed and saved Region5, and open the map document.

2. Right-click the Saudi Arabia Southern Boundaries data frame and click Activate. Expand the data frame legend.

3. Turn on the 20th Century Boundary layer. This layer reflects the boundary agreements Saudi Arabia made with most of its southern neighbors at the end of the twentieth century.

Q26 *Are the boundaries what you expected them to be?*

Q27 *Which boundary remained unsettled?*

According to international boundary expert Richard Schofield, this boundary was "the last missing fence in the desert." The only part of the boundary that was mutually agreed upon was the western area adjacent to the Red Sea. Over the years, the boundary has shifted. You will now add layers that reflect some of the major boundary changes.

 4. Click the Add Data button. Navigate to the module 5 Data folder and look in the LayerFiles folder (**OurWorld2\Mod5\Data\LayerFiles**). Hold down the Shift key and click once on each of the following file names: **Yemen1.lyr, Yemen2.lyr,** and **Yemen3 .lyr**. Click Add.

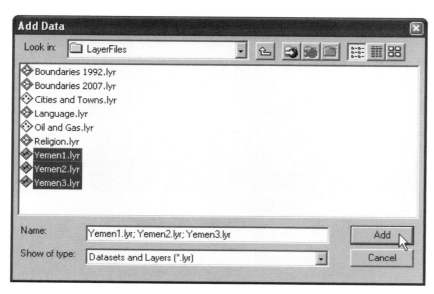

5. Turn off Yemen2 and Yemen3 for the moment and look at the red line of Yemen1.

This line represents the boundary between Yemen and Saudi Arabia established by the Treaty of Ta'if in 1934. It is the only part of the boundary that both countries recognized at the turn of the twenty-first century.

6. Turn on Yemen2.

The green line of Yemen2 represents the Saudi–Yemeni border recognized by Yemen at the end of the twentieth century. It is based on lines established when a portion of Yemen called the Aden Protectorate was under British control in the early twentieth century. Most maps used these lines to delineate the extent of Yemen prior to 2000. This boundary was not recognized by Saudi Arabia.

7. Turn on Yemen3.

    This purple line represents the Saudi–Yemeni border claimed by Saudi Arabia at the end of the twentieth century. It is based on lines established by the Saudis in the mid-1930s. This line was still being used on Saudi Arabian maps to represent the boundary in the 1990s.

8. Click the Zoom In tool. On your map, click the label Yemen. Now the map is centered on the country of Yemen.

9. Click the Fixed Zoom In button two or three times until the country of Yemen fills the view.

    (Q28) *What does the area between the green and purple lines represent?*

10. Turn on Agriculture and expand its legend.

    (Q29) *What is the principal economic activity of the regions in dispute?*

11. Turn off Agriculture and turn on and expand Population Density.

    (Q30) *Describe the population distribution in the disputed territory.*

    If you were asked to settle the boundary dispute between Saudi Arabia and Yemen, where would you draw the line? You will now draw a proposed boundary between Saudi Arabia and Yemen.

12. Make sure the Draw toolbar is displayed. If you don't see it, right-click in the gray space to the right of the Help menu to display the toolbar list and click Draw. A good place to dock the Draw toolbar is at the bottom of the ArcMap window.

13. On the Draw toolbar, click the drop-down arrow next to the New Rectangle tool and choose the New Line tool.

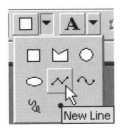

14. On your map, click the eastern end point of the red boundary line (this is the boundary that both countries agree on). Proceed eastward (to your right) and click a proposed boundary line. Double-click when you get to the end of your boundary. Now you have an additional black line that extends from the red line to Oman.

15. Make sure the Select Elements tool is now active. Click anywhere on the map away from the line you drew to make the blue selection box disappear.

Step 7: Draw the Saudi–Yemeni boundary

In June of 2000, Saudi Arabia and Yemen signed the Treaty of Jeddah, which settled their 65-year-long boundary dispute. The boundary agreement had three parts. The first part of the treaty reaffirmed agreement on the 1934 Ta'if line (the Yemen1 line). The agreement did say, however, that the line would be amended in any place where it cuts through villages.

1. Click the Zoom In tool. Zoom to the area of the Ta'if line (red line) by dragging a box around it.

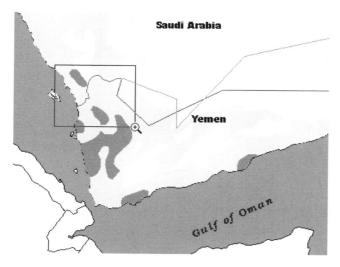

2. Click the Add Data button. Navigate to the module 5 Data folder and look in the LayerFiles folder (**OurWorld2\Mod5\Data\LayerFiles**). Select **Cities and Towns.lyr.** Click Add.

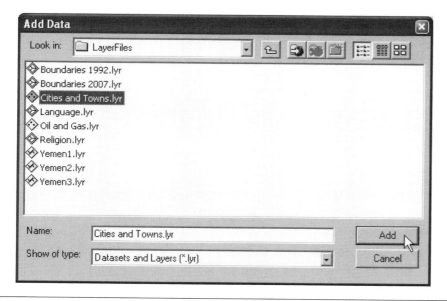

Q31 *Does the red line go through any cities or towns? (You may need to zoom in again.) If yes, approximately how many does the boundary pass through?*

Q32 *How would you decide on which side of the town to put the boundary? Remember, this decision would determine whether the residents of that village would be citizens of Saudi Arabia or Yemen.*

The second part of the Treaty of Jeddah determined the new boundary from the end of the red line to the border with Oman, 500 miles to the east. The treaty did not actually draw the line but gave its starting and ending points and points in between as latitude/longitude grid coordinates. You will now plot these points on your map to locate the new boundary line.

3. Turn off Cities and Towns and Population Density.

4. Click the View menu, point to Bookmarks, and choose Yemen to zoom out to the entire country.

You will need your own layer for holding the data you will plot. You will make a copy of the Boundary Template layer, which has no features, for this purpose.

5. Right-click the Boundary Template layer. Point to Data and click Export Data.

6. Click the Browse button in the Export Data dialog box. The Saving Data dialog box is displayed. You will be saving a feature class, so click the "Save as type" drop-down list and choose "File and Personal Geodatabase feature classes."

7. Use the "Look in" drop-down menu to navigate to the MiddleEast geodatabase (**Mod5\Data\MiddleEast.gdb**). Name the file **ABC_Yemen4**, where ABC are your initials. Click Save and then click OK in the Export Data dialog box.

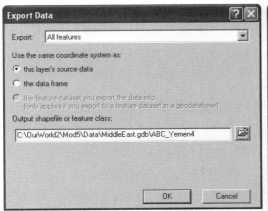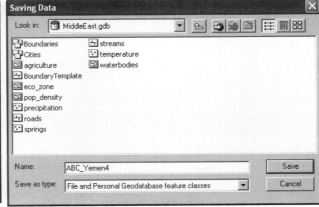

M5
L2

8. Click Yes to add the exported data to the map. Check the box to turn on ABC_Yemen4.

☐ 🖳 **Saudi Arabia Southern Boundaries**
   ☐ ☐ Cities and Towns
         •
   ☐ ☐ Major Cities
         ◆
   ☐ ☑ ABC_Yemen4
         —
   ☐ ☑ Yemen1
         —
   ☐ ☑ Yemen2
         —

9. Right-click the Boundary Template layer and click Remove to remove it from the map.

You will use tools on the Editor toolbar to plot the latitude and longitude coordinates for the new Saudi–Yemeni boundary.

10. Click the Editor Toolbar button to turn on the Editor toolbar. Dock the toolbar above the map.

11. Click the Editor menu on the Editor toolbar and choose Start Editing.

12. On the Editor toolbar make sure the Task is set to Create New Feature. Make sure the Target is your layer: ABC_Yemen4.

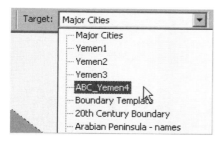

13. Click the Sketch Tool on the Editor toolbar.

14. Right-click anywhere in the map and choose Absolute X,Y.

The first point you will enter is 52 degrees longitude, 19 degrees latitude.

15. Type **52** for X. Press the Tab key and type **19** for Y. Press the Enter key.

You see a red square called a vertex appear at the border with Oman. Your cursor is attached to the vertex with an elastic line. As you move your cursor around the map without clicking, the line changes. The elastic line tells you that the new line feature you are creating is not yet finished.

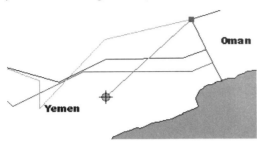

16. To enter the remaining longitude and latitude points determined in 2000, consult the table below (points 2–17). For each point, you must right-click on the map, choose Absolute X,Y, and enter the coordinates from the table. Enter all the points now.

> ↶ If you make a mistake entering a point, click the Undo button to delete it. Then enter the coordinates for that point again. If you want to delete all the points you entered and start over, double-click to complete the line and then press the Delete key. Enter the points again beginning with point number 1.

Point	Longitude	Latitude	Point	Longitude	Latitude
1	52.00	19.00	10	46.37	17.23
2	50.78	18.78	11	46.10	17.25
3	49.12	18.61	12	45.40	17.33
4	48.18	18.17	13	45.22	17.43
5	47.60	17.45	14	44.65	17.43
6	47.47	17.12	15	44.57	17.40
7	47.18	16.95	16	44.47	17.43
8	47.00	16.95	17	44.37	17.43
9	46.75	17.28			

17. When you are satisfied that you have entered all of the points correctly, right-click anywhere on the map and choose Finish Sketch. The completed line is highlighted in blue.

18. Click the Editor menu on the Editor toolbar and choose Save Edits.

(Q33) *Does the new line seem to favor Yemen or Saudi Arabia? Explain.*

Step 8:   Add the maritime portion of the boundary

The third and final part of the Treaty of Jeddah clarified the maritime boundary between Saudi Arabia and Yemen. A maritime boundary defines the offshore limits of a country. It too was defined by a series of latitude/longitude grid coordinates.

1.   Follow the procedure you just used to draw the land border to now map the maritime boundary between Saudi Arabia and Yemen.

Point	Longitude	Latitude
1	42.77	16.40
2	42.15	16.40
3	41.78	16.29

2.   Finish the sketch. Click the Editor menu and choose Stop Editing. Click Yes to save your edits.

3.   Click the Editor Toolbar button to turn off the toolbar.

4.   Click the Fixed Zoom Out button twice.

(Q34) *What body of water does the maritime boundary traverse?*

Because ArcMap randomly selects a color for a new layer, you need to change it. You will also give the layer a more descriptive name.

5.   Click the name of the ABC_Yemen4 layer two times slowly and change the name to **2000 Boundary**. Right-click the line symbol and choose a dark blue color.

(Q35) *How does the actual boundary established by the Treaty of Jeddah compare with the boundary you drew earlier (black line)?*

(Q36) *Write three observations about the boundary line created by the Treaty of Jeddah (turn on Agriculture and Population Density as needed).*

6.   Save your map document.

Step 9:   Define the pastoral area

The Treaty of Jeddah included additional provisions about the new Saudi–Yemeni boundary. One of these was the creation of a "pastoral area" on either side of the boundary. Shepherds from either Yemen or Saudi Arabia are allowed to use the pastoral area and water sources on both sides of the border according to tribal traditions. The treaty declared that the pastoral area extends 20 kilometers on either side of the border. You will now map the 20-kilometer pastoral area.

Q37 *How many miles is 20 kilometers? (1 kilometer = 0.6214 miles)*

1. Click the Selection menu and choose Set Selectable Layers.

2. Click the Clear All button in the Set Selectable layers dialog box to uncheck all of the layers. Then click the box for 2000 Boundary to make it the only layer that is checked. Click Close.

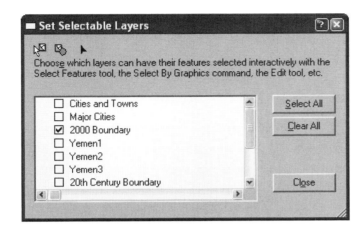

3. Click the Select Features tool on the Tools toolbar.

4. Click the blue 2000 Boundary line that crosses the land. It becomes highlighted blue on the map.

5. Click the Show/Hide ArcToolbox Window button to open the ArcToolbox window.

   ArcToolbox is where you can access many ArcGIS tools that work on your data. You will use the Buffer tool to draw the 20-kilometer zone around the boundary line.

6. Expand the Analysis Tools toolbox by clicking on the plus sign and then expand the Proximity toolbox. Double-click the Buffer tool to open the Buffer dialog box.

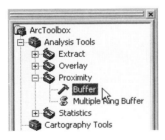

7. Click the drop-down arrow for the Input Features drop-down list and choose 2000 Boundary.

8. The default Output Feature Class location should be Mod5\Data\MiddleEast.gdb. If it is not, click the browse button and navigate to this location. Name the output feature class **ABC_Yemen4_Buffer1**.

9. For Distance, choose Kilometers from the Linear unit drop-down list, and then type **20** in the box on the left as the buffer distance.

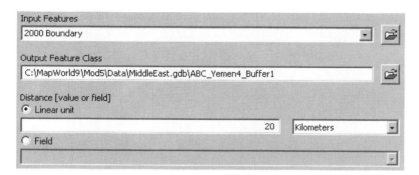

10. Click OK. After the buffer is completed, close the Buffer window if it's open. Add the buffer layer to the map if it isn't added automatically.

11. Click the Clear Selected Features button to clear the selection.

(Q38) *In which portion of the Saudi–Yemeni border will the pastoral area be most significant? Explain.*

(Q39) *Why do you think the Treaty of Jeddah created a pastoral area?*

Step 10: Create a map of the Arabian Peninsula

Before you print a map of the Arabian Peninsula, you need to clean it up.

1. Click the Select Elements button. Click the boundary line you first drew (it's black and doesn't have the buffer around it). A dashed box appears to show the line is selected. Press the Delete key. Your line disappears from the map.

2. Go to the Arabian Peninsula bookmark.

3. Decide what layers you want to display on your map. Include the following layers:
   • Major Cities
   • 2000 Boundary
   • Yemen1
   • 20th Century Boundary
   • Arabian Peninsula - names
   • Arabian Peninsula - area
   • Neighbors - countries

4. From the File menu, click Page and Print Setup. Follow your teacher's instructions to select the correct printer name. Be sure to check the "Use Printer Paper Settings" box and the "Scale Map Elements proportionally to changes in Page Size" box. In the Paper section, choose the Landscape orientation. Click OK.

5. From the File menu, click Print. Click OK.

6. Save your map document.

7. From the File menu, click Exit. When asked if you want to save your changes, click No.

In this lesson, you explored physical and population characteristics of the Arabian Peninsula. After analyzing this data, you explored boundary issues in this region and plotted the new Yemeni–Saudi boundary established by the 2000 Treaty of Jeddah.

M5
L2

Answer sheet
Module 5, Lesson 2

A line in the sand

Step 2: Identify countries that border the Arabian Peninsula

Q1)   What countries border the Arabian Peninsula to the north?

_____   _____   _____

Step 3: Investigate the physical characteristics of the Arabian Peninsula

Q2)   Is any part of the Arabian Peninsula mountainous? _____

Q3   If so, where are the mountains located?

_____

_____

Q4)   Are there any parts of the Arabian Peninsula that do not have any water at all? If so, where are these regions?

_____

_____

Q5)   Do you see any relationship between landforms and the availability of water?

_____

_____

Q6)   Describe the bodies of water in terms of their connectedness or disconnectedness.

_____

_____

Q7)   How many millimeters equal 10 inches? _____

Q8) Based on the amounts of rainfall displayed on the map, do you think there is much farming on the Arabian Peninsula? Explain.

_____

_____

_____

Q9) Approximately what percentage of the Arabian Peninsula is desert? _____

Q10) What is the approximate range of temperatures across the Arabian Peninsula during September through November?

° C: _____ ° F: _____

Q11) Which season is the hottest? _____

Q12) What is the approximate range of temperatures across the Arabian Peninsula during the hottest season?

_____ to _____ ° C _____ to _____ ° F

Q13) What relationship do you see between the Arabian Peninsula's ecozones and its patterns of landforms, precipitation, and temperature?

_____

_____

_____

Q14) List three observations for each physical characteristic in the table.

Physical characteristic	Observations
Landforms and bodies of water	
Climate	
Ecozones	

Q15) In your opinion, which of the region's physical characteristics would be considered "valuable" in a boundary decision? Explain.

_____

_____

_____

Step 4: Investigate the population characteristics of the Arabian Peninsula

Q16) What is the principal agricultural activity on the peninsula?

_____

Q17) Based on what you now know about the physical characteristics of the region, why do you think the agricultural activity is so limited?

_____

_____

_____

Q18) How does Yemen compare to the rest of the Arabian Peninsula in population density?

_____

_____

_____

Q19) Describe the overall population density of the Arabian Peninsula.

_____

_____

_____

Q20) Speculate about the ways water is most commonly used at these springs and water holes.

_____

_____

_____

Q21) Use your answers from Q16–Q20 and analysis of the maps to list two observations for each population characteristic in the table.

Population characteristics	Observations
Agricultural activities	
Population density and distribution	

Q22) If an international boundary were to be drawn across some part of the Arabian Peninsula, how would these population characteristics influence the perception of certain regions as being more valuable than others?

_____

_____

_____

Step 5: Investigate the Empty Quarter

Q23) List two observations on the physical characteristics of the Empty Quarter and two observations on its population characteristics.

Physical characteristics: _____

_____

_____

Population characteristics: _____

_____

_____

Q24) What difficulties would an area like this present if an international boundary must cross it?

_____

_____

_____

Q25) Write the map document's new name and location.

Document _____
(Example: ABC_Region5.mxd)

Location _____
(Example: C:\Student\ABC)

Step 6: Explore Saudi Arabia's southern boundaries

Q26) Are the boundaries what you expected them to be?

_____

Q27) Which boundary remained unsettled?

_____

Q28) What does the area between the green and purple lines represent?

_____

_____

Q29) What is the principal economic activity of the regions in dispute?

_____

_____

Q30) Describe the population distribution in the disputed territory.

_____

_____

_____

Step 7: Draw the Saudi–Yemeni boundary

Q31) Does the red line go through any cities or towns? (You may need to zoom in again.) If yes, approximately how many does the boundary pass through?

_____

_____

_____

Q32) How would you decide on which side of the town to put the boundary? Remember, this decision would determine whether the residents of that village would be citizens of Saudi Arabia or Yemen.

_____

_____

_____

Q33) Does the new line seem to favor Yemen or Saudi Arabia? Explain.

_____

_____

_____

Step 8: Add the maritime portion of the boundary

Q34) What body of water does the maritime boundary traverse?

_____

Q35) How does the actual boundary established by the Treaty of Jeddah compare with the boundary you drew earlier (black line)?

_____

_____

_____

_____

Q36) Write three observations about the boundary line created by the Treaty of Jeddah (turn on Agriculture and Population Density as needed).

_____

_____

_____

_____

_____

_____

Step 9: Define the pastoral area

Q37) How many miles is 20 kilometers? (1 kilometer = 0.6214 miles) _____

Q38) In which portion of the Saudi-Yemeni border will the pastoral area be most significant? Explain.

Q39) Why do you think the Treaty of Jeddah created a pastoral area?

_____

_____

_____

_____

_____

Middle school assessment
Module 5, Lesson 2

A line in the sand

You are a newspaper reporter assigned to cover the Treaty of Jeddah, signed on June 12, 2000, which settled the 65-year-old border dispute between Saudi Arabia and Yemen. Choose to be a reporter for a newspaper in either Saudi Arabia or Yemen and write your article from that country's perspective. In preparing your article, you may use the "Line in the sand" map document as well as additional resources such as your history and geography books, encyclopedias, and the Internet. Your article should include the following:

- A map showing the new boundary line and a relevant physical or cultural characteristic discussed in your article
- A description of the physical and cultural characteristics of the region affected by the boundary change
- A description of the new boundary established by the treaty and its implications for people living in the affected areas

Use the remainder of this page to brainstorm for your article.

**M5
L2**

High school assessment
Module 5, Lesson 2

A line in the sand

You are a newspaper reporter assigned to cover the Treaty of Jeddah, signed on June 12, 2000, which settled the 65-year-old border dispute between Saudi Arabia and Yemen. Choose to be a reporter for a newspaper in either Saudi Arabia or Yemen and write your article from that country's perspective. In preparing your article, you may use the "Line in the sand" map document as well as additional resources such as your history and geography books, encyclopedias, and the Internet. Your article should include the following:

- A map showing the new boundary line, the boundaries claimed by Yemen and Saudi Arabia prior to the settlement, and a relevant physical or cultural characteristic discussed in your article
- A description of the physical and cultural characteristics of the region affected by the boundary change
- A description of the historical factors that contributed to this long-standing conflict
- A description of the new boundary established by the treaty and its implications for people living in the affected areas

Use the remainder of this page to brainstorm for your article.

M5
L2

Module 6, Lesson 1

The wealth of nations

A global perspective

- Activity
- Answer sheet
- Assessment
- Assessment table

Module 6, Lesson 1

The wealth of nations

Economists generally classify a country as "developing" or "developed" by determining the percentage of gross domestic product (GDP) engaged in each of three sectors of the economy—agriculture, industry, and services. A country with a high percentage of its GDP in agriculture is categorized as developing, while a country with a high percentage of its GDP in services and industry is categorized as developed.

In this activity, you will use maps of percentages of GDP in the three sectors to explore patterns of development around the world. You will also examine two other economic indicators—energy use and GDP per capita—and compare the maps of GDP in economic sectors to the maps of GDP per capita and energy use. You will evaluate whether or not the economic sector criteria are good indicators of a country's economic status.

Step 1: Open a map document

1. Double-click the ArcMap icon on your computer's desktop.

2. When the ArcMap start-up dialog box appears, click **An existing map** and click OK.

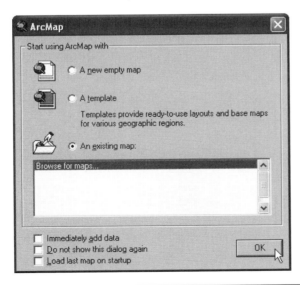

M6
L1

3. Navigate to the module 6 folder (**OurWorld2\Mod6**) and choose **Global6.mxd** (or **Global6**) from the list.

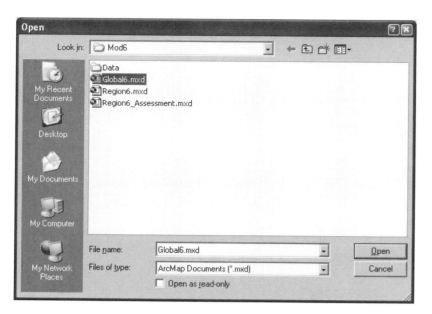

4. Click Open.

    When the map document opens, you see a world map of GDP in the agricultural sector. The table of contents lists three data frames in all, one for each of the economic sectors.

5. Click the Layout View button below the map.

    The layout shows all three data frames next to each other on the layout so you can compare them. Also, the Layout toolbar becomes active.

6. If your Layout toolbar is floating, dock it above the table of contents.

7. Maximize your ArcMap window by clicking the Maximize button in the top right corner of the window.

8. Click the Zoom Whole Page button on the Layout toolbar to make sure the layout fills the view.

Step 2: Examine the legends and patterns of the maps

GDP is defined by the *CIA World Factbook* as the "value of all final goods and services produced within a nation in a given year." The main categories are agriculture, services, and industry.

Remember to think of the percentages of GDP in agriculture, services, and industry as percentages of the total value of goods and services produced in a country and not as percentages of a country's total workforce.

1. Look at all of the legends in the table of contents.

Answers to questions in this activity should be recorded on the answer sheet.

(Q1) *What do the darkest colors represent?*

(Q2) *What do the lightest colors represent?*

2. Study the "% of GDP by sector—Agriculture" map.

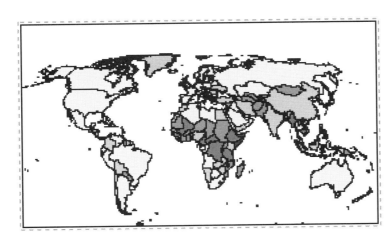

(Q3) *What is the range of percentages for the darkest color of the agriculture layer in the table of contents?*

(Q4) *Most countries with >40% of GDP in agriculture are located on which continent?*

(Q5) *On which continents do all countries have ≤40% of GDP in agriculture?*

3. Study the "% of GDP by sector—Services" map.

M6
L1

Q6 *On which continents do all countries have >40% of GDP in services?*

Q7 *Most countries with ≤40% of GDP in services are located on which continents?*

Q8 *What relationship, if any, do you see between the agriculture and services maps?*

4. Study the "% of GDP by sector—Industry" map.

Q9 *Which continent has the most countries with ≤20% of GDP in industry?*

Q10 *According to the three economic sector maps and your answers in Q4–Q9, where are most of the developing countries located?*

Step 3: Analyze data for Ecuador

1. Click the Zoom In tool on the Tools toolbar.

2. Click and drag a box around South America in the "% of GDP by sector – Industry" map.

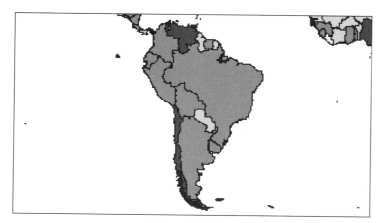

3. Zoom in on South America on the other two maps.

4. Check to see if the Agriculture data frame is active (outlined with blue dashes). If it is not, click the Select Elements tool and click the Agriculture map to activate the data frame.

5. Click the Find tool.

6. Type **Ecuador** in the Find dialog box on the Features tab.

7. Click the circle next to "In field" in the dialog box (if the button is not selected already). Then select CNTRY_NAME from the drop-down menu as the field to search.

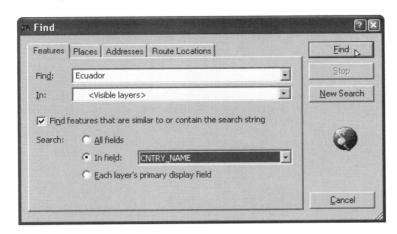

8. Click Find. Ecuador is listed in the results box at the bottom of the Find window.

9. Move the Find window so you can see the Agriculture map.

10. Right-click the Ecuador row in the results box at the bottom of the Find dialog box and choose Flash so that you can pick out Ecuador on the map.

11. Right-click Ecuador again and choose Identify. The Identify window shows data on Ecuador.

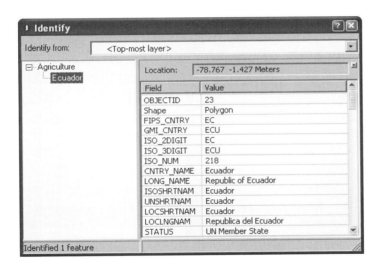

12. Scroll down the right side of the window to the rows showing percent GDP in agriculture (PGDP_AG), industry (PGDP_IN), and services (PDGP_SV).

(Q11) *What percentage of Ecuador's GDP is in agriculture?*

(Q12) *What percentage of Ecuador's GDP is in industry?*

(Q13) *What percentage of Ecuador's GDP is in services?*

Q14 *Assuming that a country can be considered "developed" if it has a high percentage of its GDP in industry and services and lower percentages of its GDP in agriculture, would you classify Ecuador as a developed or developing country? Explain.*

13. Close the Identify window and the Find window.

Q15 *Place a check mark for Ecuador under Developed or Developing in the table on the answer sheet.*

Step 4: Analyze data for other countries

1. Click the Find tool and type **Saudi Arabia** in the Find text box (if two countries are found, look for the one that has the field CNTRY_NAME). Click Find.

2. Right-click the Saudi Arabia row and choose Zoom to. The map zooms to the country of Saudi Arabia.

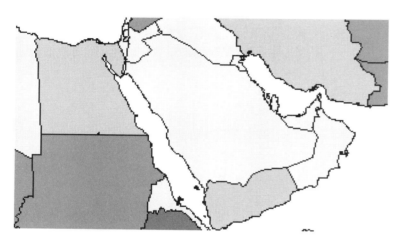

Q16 *Record the category for Saudi Arabia's percentage of GDP in agriculture in the table in Q15 using the following scale:*

Percent	Category
0 – 20	Low
20 – 40	Moderate
40 – 100	High

3. Using the Select Elements tool, click on the "% of GDP by sector – Industry" data frame to activate it.

4. Click Find in the Find dialog box to search the newly active data frame. Then Zoom to Saudi Arabia.

Q17 *Record the category for Saudi Arabia's percentage of GDP in industry in the table in Q15.*

5. Activate the "% of GDP by sector – Services" data frame. Click Find to find Saudi Arabia and then zoom in on it.

Q18 *Record the category for Saudi Arabia's percentage of GDP in services in the table in Q15.*

Q19 *Place a check mark for Saudi Arabia under Developed or Developing in the table in Q15 (refer to the criteria given in Q14).*

Q20 *Fill in the information for Australia, South Korea, Ukraine, and People's Democratic Republic of Congo (Congo, DRC) in the table in Q15 (repeat steps 1–5 for each country).*

6. Close the Find dialog box.

7. Use the Select Elements tool and the Full Extent button to activate each data frame and zoom to the entire world in all three maps.

Step 5: Create a new data frame and add data

You will now add GDP per capita data and energy use data and determine whether this data supports your initial conclusions about which countries are developed and which are developing.

Remember, GDP is the total value of all goods, services, and products produced in a given country. Typically, developing countries have a low GDP per capita. The total amount of energy consumed in a given country is also an indicator of development. If a country has a low level of energy consumption, it tends to be a developing country. Developed countries are high in both GDP per capita and energy use.

1. Click Insert on the Main Menu toolbar and choose Data Frame to add a new data frame to the map document.

2. With the Select Elements tool, click the new data frame and drag it into the blank space in the bottom right of the layout.

3. Click and drag the blue boxes to resize the data frame to match the others.

**M6
L1**

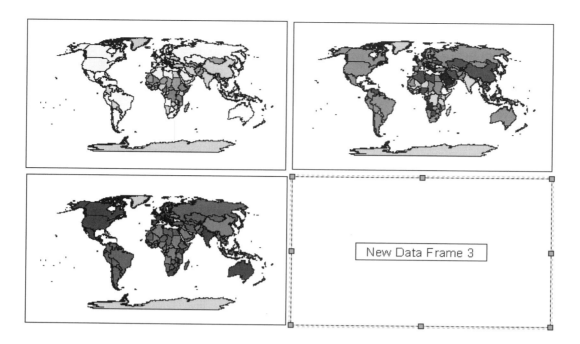

4. In the table of contents, click slowly two times on New Data Frame 3 to activate the text cursor. Name the data frame **Economic Data**. (If the Data Frame Properties dialog box opens instead, click the General tab. Type **Economic Data** in the Name box, then click OK.)

5. Click the Add Data button.

6. Navigate to the module 6 Data folder (**OurWorld2\Mod6\Data**). Open the World6. gdb geodatabase Press the Ctrl key and select both **cntry04_energy** and **cntry07_ econ**. Click Add.

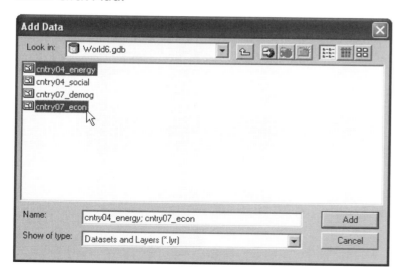

7. Turn off cntry04_energy in the table of contents.

For each new layer, all countries are symbolized with one color. ArcMap randomly assigns the color, so the color on your screen may not match the color on your neighbor's screen.

Step 6:   Thematically map GDP per capita and energy use

GDP per capita is the gross domestic product per person in a given year (we will use the data for 2007). It is calculated by dividing the GDP of a country by the country's total population.

1. Double-click the cntry07_econ layer in the table of contents. The Layer Properties dialog box is displayed.

2. Click the Symbology tab.

3. Click Quantities in the Show list on the left side of the dialog box. The "Graduated colors" option is automatically selected.

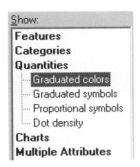

4. Click the Value drop-down menu, scroll down, and select GDP_PCAP. A list of five colored symbols, values, and labels appears.

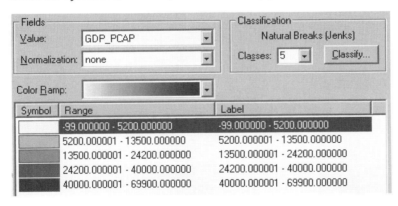

5. For the Color Ramp choose Brown Light to Dark. To choose the color ramps by name instead of colors, right-click the drop-down menu and choose Graphic View to toggle the colors off.

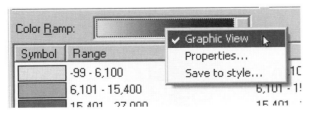

By default, ArcMap has divided the GDP per capita values into five groups, or classes, which are shown in the Range column.

For some countries no data is available for GDP per capita. For those countries, the GDP_PCAP field has a value of –99. Because the value of –99 represents no data, it should not be included in the range of GDP per capita values. You will assign values of –99 to a No Data class.

6. Click the Classify button on the right. Click the Exclusion button in the Classification dialog box.

7. Click the Query tab (if it's not already selected) in the Data Exclusion Properties dialog box and click the Clear button to clear any existing queries (if there are no existing entries to clear, the Clear button is grayed out).

You will now build the following query expression: [GDP_PCAP] = –99

8. Scroll down the "Exclude clause" list and find GDP_PCAP. Double-click GDP_PCAP to add it to the expression box.

9. Click the equals button (=).

10. Click the Get Unique Values button and then double-click –99 to complete the query expression.

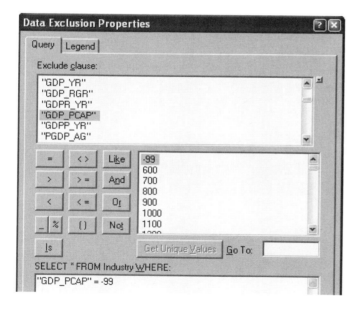

11. Click the Verify button. If you receive a message that the expression was successfully verified, click OK to close the message. If you get a different message, check your work and make sure the query expression matches the one pictured above.

12. Click the Legend tab at the top of the dialog box. Check the box to "Show symbol for excluded data."

13. Click Symbol color to open the Symbol Selector. Click the Fill Color symbol and change the fill color to Gray 20% (you can see the color name and percent if you hover over the color boxes). Click OK.

14. In the Label box, type No Data.

15. Click OK in the Data Exclusion Properties and Classification dialog boxes.

    The ranges in the Symbology tab no longer include the value –99.

    Next you will change the labels for each symbol to whole numbers (without decimals).

16. Under Label, right-click on any number and choose Format Labels from the list.

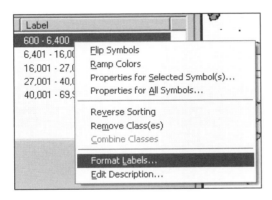

17. Reduce the number of decimal places to zero in the Number Format dialog box. Check Show thousands separators. Click OK.

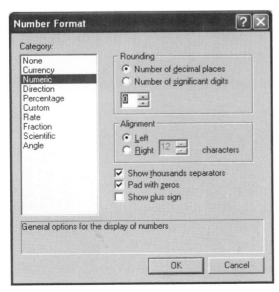

M6
L1

18. Click the General tab in the Layer Properties window. Change the Layer Name to GDP per capita ($).

19. Click OK and look at the updated legend.

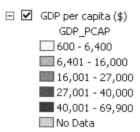

Next you will map energy use, which is the total annual energy consumption from all energy sources as measured in quadrillion BTUs (British thermal units). One BTU is the amount of energy needed to heat one pound of water one degree Fahrenheit.

20. Turn off GDP per capita and turn on cntry04_energy.

21. Double-click the cntry04_energy layer to open its Layer Properties dialog box. Repeat steps 2–19 using the information below:
- Use graduated colors to map the value field ENER_USE.
- Use the Yellow to Dark Red color ramp.
- In Classify/Exclusion, use the query expression [ENER_USE = -99].
- Change the layer name to **Energy Use (quadrillion BTUs)**.

Step 7: Analyze GDP per capita and energy use data

Now you will take this new data on GDP per capita and energy use into consideration, and you will reevaluate how you classified countries as developed or developing.

1. Turn off Energy Use and turn on GDP per capita.

2. Click the Find tool and find Ecuador. Zoom to Ecuador.

(Q21) *What is Ecuador's GDP per capita?*

3. Turn off GDP per capita and turn on Energy Use.

(Q22) *What is Ecuador's annual amount of energy use?*

(Q23) *Based on this new information, should Ecuador be classified as a developing or developed country?*

(Q24) *Why does energy use increase when a country develops?*

You will now look at GDP per capita and energy use data for the other countries that you earlier classified as developing or developed.

(Q25) *Complete the table on the answer sheet (repeat items 1–3 for each country). Categorize countries with <$16,000 as low GDP/capita and countries with <11 quadrillion BTUs as low on energy use.*

(Q26) *Name a country that you classified in Q15 as developed on the basis of economic sector data and in Q25 as developing on the basis of GDP per capita and energy use data.*

(Q27) *Based on the data you collected on these six countries, do you feel that the economic sector criteria are good indicators of a country's economic status? Explain your answer.*

4. Click the Full Extent button so you can see the entire world in the Economic Data map.

5. Choose Save As from the File menu. Ask your teacher for instructions on where to save the map document and how to rename it. You will need this map document for the assessment.

(Q28) *Write the new name you gave the map document and where you saved it.*

6. Click the Restore Down button to return the ArcMap window to its original size.

7. From the File menu, click Exit.

In this lesson, you used economic sector data to determine whether countries should be classified as developed or developing. You added new data, thematically mapped the data, and reevaluated your previous classifications.

**M6
L1**

Module 6, Lesson 1

The wealth of nations

Step 2: Examine the legends and patterns of the maps

Q1) What do the darkest colors represent?

_____

Q2) What do the lightest colors represent?

_____

Q3) What is the range of percentages for the darkest color of the agriculture layer in the table of contents?

_____

Q4) Most countries with >40% of GDP in agriculture are located on which continent?

_____

Q5) On which continents do all countries have ≤40% of GDP in agriculture?

_____

Q6) On which continents do all countries have >40 % of GDP in services?

_____

Q7) Most countries with ≤40 % of GDP in services are located on which continents?

_____

Q8) What relationship, if any, do you see between the agriculture and services maps?

_____

Q9) Which continent has the most countries with ≤20% of GDP in industry?

_____

Q10) According to the three economic sector maps and your answers in Q4–Q9, where are most of the developing countries located?

_____

Step 3: Analyze data for Ecuador

Q11) What percentage of Ecuador's GDP is in agriculture? _____

Q12) What percentage of Ecuador's GDP is in industry? _____

Q13) What percentage of Ecuador's GDP is in services? _____

Q14) Would you classify Ecuador as a developed or developing country? Explain.

_____

_____

_____

_____

Q15) Place a check mark for Ecuador under Developed or Developing in the table.

Country	GDP in the following sector:			Economic status	
	Agriculture	**Industry**	**Services**	**Developing**	**Developed**
Ecuador	Low	Moderate	High		
Saudi Arabia					
Australia					
South Korea					
Ukraine					
Congo DRC					

Step 4: Analyze data for other countries

Q16) Record the category for Saudi Arabia's percentage of GDP in agriculture in the table in Q15.

Q17) Record the category for Saudi Arabia's percentage of GDP in industry in the table in Q15.

Q18) Record the category for Saudi Arabia's percentage of GDP in services in the table in Q15.

Q19) Place a check mark for Saudi Arabia under Developed or Developing in the table in Q15.

Q20) Fill in the information for Australia, South Korea, Ukraine, and People's Democratic Republic of Congo (Congo, DRC) in the table in Q15.

Step 7: Analyze GDP per capita and energy use data

Q21) What is Ecuador's GDP per capita? _____

Q22) What is Ecuador's annual amount of energy use? _____

Q23) Based on this new information, should Ecuador be classified as a developing or developed country?

_____

Q24) Why does energy use increase when a country develops?

_____

_____

_____

_____

Q25) Complete the following table. Categorize countries with <$16,000 as low GDP/capita and countries with <11 quadrillion BTUs as low on energy use.

Country	GDP per capita ($)	Energy use (quadrillion BTUs)	Developed or developing	Is this a change from your earlier classification?
Ecuador				
Saudi Arabia				
Australia				
Republic of Korea (South Korea)				
Ukraine				
Congo DRC				

Q26) Name a country that you classified in Q15 as developed on the basis of economic sector data and in Q25 as developing on the basis of GDP per capita and energy use data.

_____

Q27) Based on the data you collected on these six countries, do you feel that the economic sector criteria are good indicators of a country's economic status? Explain your answer.

_____

_____

_____

_____

_____

_____

_____

_____

_____

Q28) Write the new name you gave the map document and where you saved it.

Document _____
(Example: ABC_Global5.mxd)

Location _____
(Example: C:\Student\ABC)

Middle school assessment
Module 6, Lesson 1

The wealth of nations

1. Open your saved version of the Global6 map document (e.g., ABC_Global6). Use the three "% of GDP by sector" maps (agriculture, services, and industry) to select a country you believe is a developed country and another country that you believe is a developing country. Write the names of those countries at the top of the assessment table.
2. Thematically map at least eight of the following indicators of economic development from the attribute tables in the World6 geodatabase.

Indicator	Field name	Indicator	Field name
Attribute table: cntry07_demog			
Birth rate	CBR2007	Life expectancy	LE2007
Death rate	CDR2007	Population under 15	POP014_07
Infant mortality rate	IMR2007	Urban population	PCT_URBAN
Attribute table: cntry04_social			
Literacy	LITR_2007		
Attribute table: cntry07_econ			
GDP per capita (already mapped)	GDP_PCAP	Phones per person	PHONLNS (Value field) with POP2007 (Normalization field)
GDP per economic sector (already mapped)	PGDP_AG, PGDP_IN, PGDP_SV		
Attribute table: cntry04_energy			
Energy consumption (already mapped)	ENER_USE		

To begin, activate the Economic Data data frame and add your first layer from the World6 geodatabase. Switch to Data View.

You will need to add other layers into the data frame for each attribute you map. You may want to collapse any unneeded data frames in the table of contents.

M6 L1

3. Place a check mark in the assessment table for any indicator of low economic development you find to be present in either of the two countries you selected.
4. Look at the data in the two columns and decide if your initial prediction was correct.
5. Write an essay that discusses the following:
   - The three economic sectors
   - How well the data for your two countries supports the assertion that GDP per economic sector is an indicator of economic development ("developed" or "developing")
   - Your own definitions, based on your research, of a developed country and a developing country

High school assessment
Module 6, Lesson 1

The wealth of nations

1. Open your saved version of the Global6 map document (e.g., ABC_Global6). Use the three "% of GDP by sector" maps (agriculture, services, and industry) to select a country you believe is a developed country and another country that you believe is a developing country. Write the names of those countries at the top of the assessment table.
2. Thematically map the following indicators of economic development from the attribute tables in the World6 geodatabase. You may use additional indicators from any of the tables as well.

Indicator	Field name	Indicator	Field name
Attribute table: cntry07_demog			
Birth rate	CBR2007	Life expectancy	LE2007
Death rate	CDR2007	Population under 15	POP014_07
Infant mortality rate	IMR2007	Urban population	PCT_URBAN
Attribute table: cntry04_social			
Literacy	LITR_2007		
Attribute table: cntry07_econ			
GDP per capita (already mapped)	GDP_PCAP	Phones per person	PHONLNS (Value field) with POP2007 (Normalization field)
GDP per economic sector (already mapped)	PGDP_AG, PGDP_IN, PGDP_SV		
Attribute table: cntry04_energy			
Energy consumption (already mapped)	ENER_USE		

M6
L1

To begin, activate the Economic Data data frame and add your first layer from the World6 geodatabase. Switch to Data View.

You will need to add other layers into the data frame for each attribute you map. You may want to collapse any unneeded data frames in the table of contents.

3. Place a check mark in the assessment table for any indicator of low economic development you find to be present in either of the two countries you selected.
4. Look at the data in the two columns and decide if your initial prediction was correct.
5. Write an essay that discusses the following:
    *   The relationship between the three economic sectors (for example, will a country usually have high percentages or low percentages in all three production areas?)
    *   How well the data for your two countries supports the assertion that GDP per economic sector is an indicator of economic development ("developed" or "developing")
    *   Your own definitions, based on your research, of a developed country and a developing country
    *   Prediction of changes in your selected countries' economic statuses by the year 2020 (include supporting data)

Assessment table

The wealth of nations

Indicator of low economic development	Country believed to be developed:	Country believed to be developing:
High birth rate		
High death rate		
High infant mortality rate		
Low life expectancy at birth		
Large percentage of population under 15		
Low literacy rate		
High proportion of population living in urban areas		
Low GDP per capita		
Low consumption of energy		
Low number of phones per person		
High GDP percentage in agriculture, low GDP percentages in industry and services		
Other:		
Other:		
Other:		

M6
L1

Module 6, Lesson 2

Share and share alike

A regional investigation of North America

- Supplement: NAFTA objectives
- Activity
- Answer sheet
- Assessment

Supplement

NAFTA objectives

The North American Free Trade Agreement (NAFTA) came into effect on January 1, 1994. This agreement created the world's largest free trade area. Among the agreement's main objectives are the liberalization of trade between Canada, Mexico, and the United States; stimulation of economic growth in all three countries; and equal access to each other's markets.

Article 101. Establishment of the Free Trade Area

The Parties to this Agreement, consistent with Article XXIV of the General Agreement on Tariffs and Trade, hereby establish a free trade area.

Article 102. Objectives

1. The objectives of this Agreement, as elaborated more specifically through its principles and rules, including national treatment, most-favored-nation treatment and transparency, are to:
   a. eliminate barriers to trade in, and facilitate the cross-border movement of, goods and services between the territories of the Parties;
   b. promote conditions of fair competition in the free trade area;
   c. increase substantially investment opportunities in the territories of the Parties;
   d. provide adequate and effective protection and enforcement of intellectual property rights in each Party's territory;
   e. create effective procedures for the implementation and application of this Agreement, for its joint administration and the resolution of disputes; and
   f. establish a framework for further trilateral, regional and multilateral cooperation to expand and enhance the benefits of this Agreement.
2. The Parties shall interpret and apply the provisions of this Agreement in the light of its objectives set out in paragraph 1 and in accordance with applicable rules of international law.

From the North American Free Trade Agreement between the Government of Canada, the Government of the United Mexican States, and the Government of the United States of America, published January 1, 1994.

**M6
L2**

Module 6, Lesson 2

Share and share alike

On January 1, 1994, the North American Free Trade Agreement (NAFTA) was enacted to enhance trade and increase access to the total trade market available to businesses in North America. Tariffs and quotas were eliminated to increase the competitiveness of goods produced by all North Americans. Now, more than a decade after NAFTA's inception, your task is to evaluate whether these objectives have been accomplished for all three countries involved. In this activity you will analyze graphs of exports and trade balances for the NAFTA trading partners. Ask yourself the following questions as you are working through the activity:

- Have export levels or trade balances changed since the inception of NAFTA?
- Is the market available for business in North America larger than before?
- Is there greater competition forcing businesses to provide the best products and best prices for their goods?
- Has NAFTA been equally beneficial for all countries involved?
- Does the trade balance of a country tell the whole story of how effective NAFTA is?
- Can you tell a difference between trade in the years directly before NAFTA (1991–1993) and trade after the inception of NAFTA (1994)?

Ultimately, you will have to decide if you think NAFTA is effective!

Step 1: Open a map document

1. Double-click the ArcMap icon on your computer's desktop.

M6
L2

2. When the ArcMap start-up dialog box appears, click **An existing map** and click OK.

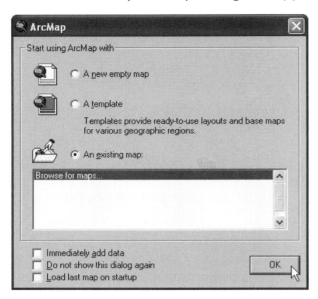

3. Navigate to the module 6 folder (**OurWorld2\Mod6**) and choose **Region6.mxd** (or **Region6**) from the list.

4. Click Open. The map of North America shows the three countries participating in NAFTA.

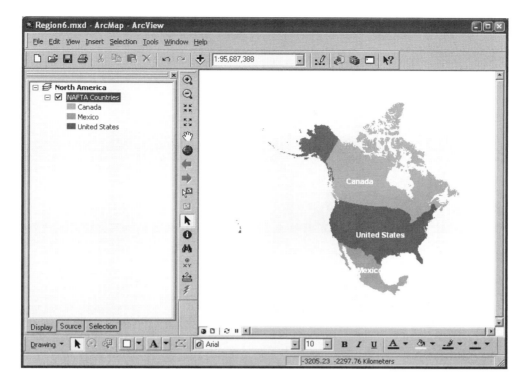

Step 2: Examine the map and attribute table

1. Click the Identify tool and then click Canada on the map.

2. Look at the attribute names in the Field column of the Identify window.

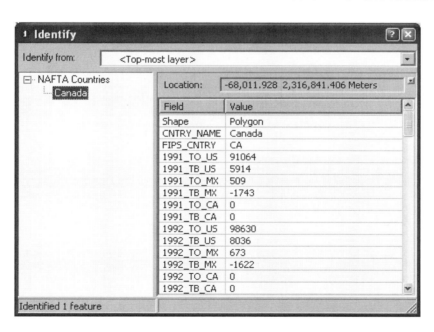

3. Scroll down the list and answer the following questions.

> Answers to questions in this activity should be recorded on the answer sheet.

M6
L2

Q1 *For which years does the layer contain data?*

Q2 *How many attributes are there for each year?*

Field names with the word "to" in them represent the value of goods and services exported from one country to another. For example, the attribute 1991 to US represents the value of goods and services that a country exported to the United States in the year 1991.

Field names with the abbreviation "TB" represent a country's trade balance, which is the difference between how much a country exports to a trading partner and how much it imports from that partner.

The values are expressed in millions of U.S. dollars. That means you have to multiply the number you see by 1,000,000 to get the actual dollar value.

Q3 *What was the value of goods and services exported from Canada to the United States in 1991?*

4.   Close the Identify window.

5.   Right-click NAFTA Countries in the table of contents and choose Open Attribute Table.

Q4 *What is the name of the table?*

Q5 *How many rows are there for each country on the map?*

6.   Click the gray box at the beginning of the United States row in the table.

⊞ Attributes of NAFTA Countries

Shape *	CNTRY_NAME	FIPS_CNTRY	1991_TO_US
▶ Polygon	Mexico	MX	31130
Polygon	United States	US	0
Polygon	Canada	CA	91064

Step 3:   Relate another table to the layer table

You will now open a second table that contains the trade data organized differently. You will then relate the two tables together and view the trading statistics in the tables and on a series of graphs.

1.   Click the Source tab at the bottom of the table of contents (if the Source tab is covered by the attribute table, click the table's title bar and drag the table out of the way).

2. Right-click NAFTA_Trading_Statistics in the table of contents and choose Open.

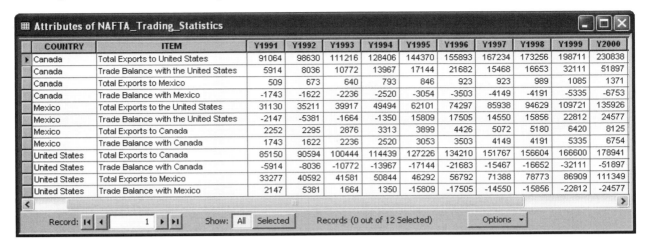

COUNTRY	ITEM	Y1991	Y1992	Y1993	Y1994	Y1995	Y1996	Y1997	Y1998	Y1999	Y2000
Canada	Total Exports to United States	91064	98630	111216	128406	144370	155893	167234	173256	198711	230838
Canada	Trade Balance with the United States	5914	8036	10772	13967	17144	21682	15468	16653	32111	51897
Canada	Total Exports to Mexico	509	673	640	793	846	923	923	989	1085	1371
Canada	Trade Balance with Mexico	-1743	-1622	-2236	-2520	-3054	-3503	-4149	-4191	-5335	-6753
Mexico	Total Exports to the United States	31130	35211	39917	49494	62101	74297	85938	94629	109721	135926
Mexico	Trade Balance with the United States	-2147	-5381	-1664	-1350	15809	17505	14550	15856	22812	24577
Mexico	Total Exports to Canada	2252	2295	2876	3313	3899	4426	5072	5180	6420	8125
Mexico	Trade Balance with Canada	1743	1622	2236	2520	3053	3503	4149	4191	5335	6754
United States	Total Exports to Canada	85150	90594	100444	114439	127226	134210	151767	156604	166600	178941
United States	Trade Balance with Canada	-5914	-8036	-10772	-13967	-17144	-21683	-15467	-16652	-32111	-51897
United States	Total Exports to Mexico	33277	40592	41581	50844	46292	56792	71388	78773	86909	111349
United States	Trade Balance with Mexico	2147	5381	1664	1350	-15809	-17505	-14550	-15856	-22812	-24577

The Attributes of NAFTA_Trading _Statistics table is not listed on the Display tab because it is not a map layer. It is listed only on the Source tab.

3. Make each table window smaller. Make sure that you can see all of the rows in the table as well as the Options button at the bottom. Arrange your windows so that you can see the map and both of the tables at the same time (your ArcMap window can be small for this investigation).

4. Examine the Attributes of NAFTA _Trading _Statistics table on your screen. Scroll the table if necessary to answer the following questions.

Q6) *How many rows are there for each country?*

Q7) *What do the rows for Canada say under Item?*

Q8) *What are the general types (or items) of information listed in the table?*

Q9) *How many years of data are represented in the table?*

You will now relate the two tables by using information they have in common (country names).

5. Right-click NAFTA Countries in the table of contents, point to Joins and Relates, and click Relate.

M6
L2

6. In the first drop-down menu in the Relate dialog box, scroll down and choose CNTRY_NAME. Keep the default selections for items 2 and 3. In item 4, replace the default name with **Countries to Years**. Click OK.

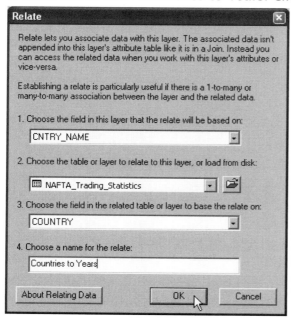

7. Click the Options button in the Attributes of NAFTA Countries table and choose Related Tables and then Countries to Years: NAFTA_Trading_Statistics.

(Q10) *What happens in the Attributes of NAFTA_Trading_Statistics table?*

8. Use the Select Features tool to click Canada on the map.

(Q11) *What happens in the two tables and the map?*

9. Update the related rows by clicking the Options button in the Attributes of NAFTA Countries table and choosing Related Tables and then Countries to Years: NAFTA_Trading_Statistics.

Next, you'll see what happens when you make a selection in the related table (Attributes of NAFTA _Trading_Statistics).

10. Select a row for Mexico in the Attributes of NAFTA _Trading _Statistics table.

(Q12) *What happens in the two tables and the map?*

11. Now click the Options button in the related table (Attributes of NAFTA_Trading_Statistics) and choose Related Tables and then Countries to Years: NAFTA_countries.

(Q13) *What have you observed about the way the NAFTA_Trading_Statistics table is tied to the NAFTA Countries table and map layer?*

12. Click the Clear Selected Features button. All of the selections are cleared.

Step 4:   Examine export graphs

1.   Click the Tools menu, point to Graphs, and choose Exports to Canada.

The graph shows three sets of data, one for each country. If you see only one country's data on the graph, click the Clear Selected Features button.

2.   Enlarge the graph window slightly so you can read it more easily.

(Q14)   *Which country exported more goods and services to Canada: Mexico or the United States?*

(Q15)   *Why is the graph empty in the space for Canada?*

3.   Drag the graph window completely off the map and table windows.

4.   Using the Select Features tool, click Mexico on the map. Mexico is outlined in blue.

(Q16)   *What happened to the graph?*

Because the graph is created from the data in the attribute table, it is tied to the map through the table. When you select a country on the map, only the data for that associated row in the table is displayed in the graph.

(Q17)   *How many years of data are represented on the graph?*

(Q18)   *What year does the first bar on the left represent?*

(Q19)   *Compare the numbers on the y-axis with those in the two tables. Are the numbers on the graph in thousands, millions, or billions of dollars? (Remember that the Attributes of NAFTA Countries table values are in millions of dollars.)*

Q20 *Looking at the graph, how would you describe the trend of Mexican exports to Canada over the 16-year period?*

Q21 *What was the approximate value of Mexican exports to Canada in 1991? In 2006?*

Q22 *Approximately how many times greater is the 2006 amount of Mexican exports to Canada than the 1991 amount?*

5. Close the Exports to Canada graph.

6. Click the Tools menu and open the graph entitled Exports to U.S.

Q23 *How would you describe the trend of Mexican exports to the United States over the 16-year period?*

Q24 *Approximately how many times greater is the 2006 amount of Mexican exports to the United States than the 1991 amount?*

Note that the same column height on two different graphs may represent different dollar values.

Step 5: Examine a trade balance graph

1. Click the Tools menu and open the graph entitled Trade Balance with U.S.

2. This graph may cover up the first graph. Drag it below the first one so you can see both graphs at the same time.

Remember, trade balance is the difference between how much a country exports to a trading partner and how much it imports from the same partner.

Q25 *Did Mexico have a trade surplus or deficit with the United States for 1992?*

Q26 *What was the approximate value of the trade balance for 1992? (Remember, the y-axis is in millions of dollars.)*

Q27 *What was the first year that Mexico exported more to the United States than it imported from the United States?*

Q28 *Describe the trend of Mexico's trade balance with the United States over the 16-year period.*

3. Click Canada on the map. Look at the Trade Balance graph again.

Q29 *Did Canada have a trade deficit with the United States anytime during the 16-year period?*

4. Click the Clear Selected Features button. Compare Mexico and Canada on the graph.

Q30 *In 1998, was Canada's trade balance with the United States greater, smaller, or about the same as Mexico's?*

Q31 *In 2000, was Canada's trade balance with the United States greater, smaller, or about the same as Mexico's?*

5. Select Canada on the map, click the Options menu in the Attributes of NAFTA Countries table, and update the related table.

Q32 *Referring to the Attributes of NAFTA_Trading_Statistics table, what was the exact value of Canada's trade balance with the United States in 2006?*

6. Click the Clear Selected Features button.

7. Close the Attributes of NAFTA Countries and Attributes of NAFTA_Trading_Statistics tables.

8. Close the two graph windows.

9. Ask your teacher if you should stop here and save this map document. Follow your teacher's instructions on how to rename the document and where to save it. (If you don't need to save the document, skip Q33.)

Q33 *Write the map document's new name and location.*

Step 6: Evaluate the effectiveness of NAFTA

In this lesson, you are looking at the level of trade among the NAFTA trading partners: Canada, Mexico, and the United States. You are also looking at changes in the level of trade among the three countries over time.

Remember, it is nearly impossible for any country to produce everything it needs to support its people. Countries must import (purchase) goods from other countries. One country's imports are its trade partner's exports. In this map document, you have no graphs for imports because you can figure out that information by looking at the export graphs.

For example, to find the United States' imports from Canada, you switch the question in your mind and look for Canada's exports to the United States. You would find this information on the graph entitled Exports to United States.

M6
L2

To decide whether or not NAFTA is meeting its goals, you will need to further explore the data in the graphs and tables in the map document. You will gather the information that you need to decide if you think NAFTA is effective. You may want to refer to the NAFTA Objectives handout or the list of NAFTA goals you developed with your teacher at the beginning of the lesson. Remember these formulas as you examine the graphs and answer the questions that follow:

Trade Balance = Exports – Imports
Trade Surplus = Exports > Imports
Trade Deficit = Imports > Exports

1. Go to the Tools menu and open the three export graphs. Arrange them so they are not overlapping and you can compare them.

(Q34) *Determine from the graphs the approximate value of exports in 2006 for each pair of countries in the table on the answer sheet. Record the values in the middle column.*

(Q35) *Add the export values together for each pair of countries (for example, exports from the United States to Mexico plus exports from Mexico to the United States). Record the totals in the last column in the table in Q34.*

(Q36) *Rank the trading partners by total volume of trade. Use 1 for the partners trading the most and 3 for the partners trading the least.*

(Q37) *Did NAFTA have a positive (+), negative (–), or neutral (N) effect on trade volume between partner countries?*

(Q38) *Do you think that any one of these three countries benefited more than the other two from NAFTA? If so, which country? Explain your answer.*

2. Close the three graph windows.

3. Go to the Tools menu and open the three Trade Balance graphs. Arrange them so you can compare them.

Most countries prefer to export a higher volume and dollar value of goods than they import, but either a large trade surplus or a large trade deficit may negatively affect a country's overall economic health.

4. Look at the Trade Balance with Canada graph.

(Q39) *What country has a healthier trade balance with Canada: Mexico or the United States?*

(Q40) *On what graph do you find a set of bars that looks like a mirror image of those for the U.S. trade balance with Canada?*

5. Compare the trade balance graphs.

(Q41) *What country had the most dramatic change for the better in trade after NAFTA came into being in 1994?*

(Q42) *Estimate U.S. trade deficits with Canada and Mexico for 2004 and 2006 and record them (in billion dollars) in the table on the answer sheet.*

(Q43) *Did the U.S. combined trade balance get better or worse between 2004 and 2006? By how much?*

6. Close the three graph windows.

Step 7: Create and print a layout

You will now create a layout presenting information about U.S. trade with Canada. The layout will include the map of NAFTA countries and two graphs: Exports to Canada and Trade Balance with Canada.

1. Click the Layout View button.

2. Maximize your ArcMap window so that it fills your screen. If your Layout toolbar is floating, dock it above the view or in your preferred location.

3. Click the Zoom Whole Page button on the Layout toolbar to make the layout fill the view.

The layout looks like a piece of paper. The layout already has a map, title, north arrow, legend, and scale bar. Before adding the graphs, you will make space for them by slightly reducing the size of the map.

4. Click the Fixed Zoom Out button on the Tools toolbar once to reduce the size of North America.

Next you will add the two graphs.

5. Click the Tools menu, point to Graphs, and then click Manage to display the Graph Manager. Right-click on Exports to Canada and click Add to layout.

The graph is added to the middle of the layout.

6. Repeat the procedure to add the Trade Balance with Canada graph to the layout.

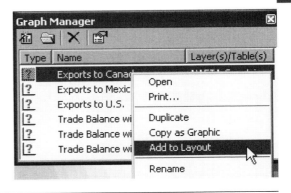

M6
L2

7. Close the Graph Manager window.

You will arrange the two graphs on the left side of the layout. You will also make the map a little smaller so that it better fits the space on the right.

8. Click the Trade Balance with Canada graph using the Select Elements tool. Drag the graph to the upper left of the layout, under the title.

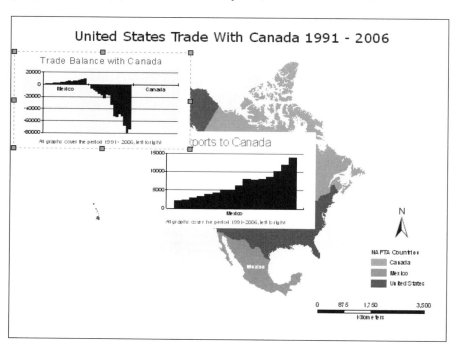

9. Click the Exports to Canada graph to activate it. Then click and drag it to a position below the first graph.

10. Use the Pan tool on the Tools toolbar to move North America slightly to the right—center it between the graphs and the legend. Make adjustments as needed until you are satisfied with the look of your layout.

You will now add your name and the date to the layout.

11. Click the New Text tool on the Draw toolbar. Click the yellow space underneath the graphs to insert text.

12. Type your name, a comma, and today's date, and then press the Enter key on your keyboard (if the text box doesn't let you type anything in, double-click on it and type the information in the Properties dialog box).

Alberto Chavez, December 10, 2008

13. Drag the text down to the lower left corner of the layout, just inside the inner frame of the layout. When you are finished, click anywhere in the white space outside the layout to unselect the text.

Finally, you need to make the graphs omit extraneous data and show only U.S. trade with Canada.

14. Click the Select Features tool and click the United States on the map. The graphs change to show only U.S. data.

15. From the File menu, select Print. Click the Setup button in the upper right corner of the dialog box. In the Page and Print Setup dialog box, make sure the Landscape orientation is selected.

16. Check with your teacher to be sure the correct printer is selected, and change it if necessary. Click OK in both dialog boxes to print your layout.

17. Ask your teacher where to save this map document and how to rename it. If you are not going to save the document, exit ArcMap by choosing Exit from the File menu. When asked if you want to save changes to the map document, click No.

In this activity, you used ArcMap tables, graphs, and a map to explore trade data for the three North American countries. After analyzing this data, you created a layout to present information on U.S. trade with Canada.

M6
L2

Name_____   Date_____

Module 6, Lesson 2

Share and share alike

Step 2: Examine the map and attribute table

Q1)   For which years does the layer contain data? _____ to _____

Q2)   How many attributes are there for each year? _____

Q3)   What was the value of goods and services exported from Canada to the United States in 1991?

       $_____

Q4)   What is the name of the table? _____

Q5)   How many rows are there for each country on the map? _____

Step 3: Relate another table to the layer table

Q6)   How many rows are there for each country? _____

Q7)   What do the rows for Canada say under Item?

_____

_____

_____

_____

Q8)   What are the general types (or items) of information listed in the table?

_____

_____

Q9)   How many years of data are represented in the table? _____

Q10) What happens in the Attributes of NAFTA_Trading_Statistics table?

_____

_____

Q11) What happens in the two tables and the map?

_____

_____

Q12) What happens in the two tables and the map?

_____

_____

Q13) What have you observed about the way the NAFTA_Trading_Statistics table is tied to the NAFTA Countries table and map layer?

_____

_____

Step 4: Examine export graphs

Q14) Which country exported more goods and services to Canada: Mexico or the United States?

_____

Q15) Why is the graph empty in the space for Canada?

_____

_____

Q16) What happened to the graph?

_____

_____

Q17) How many years of data are represented on the graph? _____

Q18) What year does the first bar on the left represent? _____

Q19) Compare the numbers on the y-axis with those in the two tables. Are the numbers on the graph in thousands, millions, or billions of dollars?

_____

Q20) Looking at the graph, how would you describe the trend of Mexican exports to Canada over the 16-year period?

_____

Q21) What was the approximate value of Mexican exports to Canada in 1991?

_____

In 2006? _____

Q22) Approximately how many times greater is the 2006 amount of Mexican exports to Canada than the 1991 amount?

_____

Q23) How would you describe the trend of Mexican exports to the United States over the 16-year period?

_____

Q24) Approximately how many times greater is the 2006 amount of Mexican exports to the United States than the 1991 amount?

_____

Step 5: Examine a trade balance graph

Q25) Did Mexico have a trade surplus or deficit with the United States for 1992?

_____

Q26) What was the approximate value of the trade balance for 1992?

_____

Q27) What was the first year that Mexico exported more to the United States than it imported from the United States?

_____

Q28) Describe the trend of Mexico's trade balance with the United States over the 16-year period.

_____

_____

Q29) Did Canada have a deficit trade balance with the United States anytime during the 16-year period?

_____

Q30) In 1998, was Canada's trade balance with the United States greater, smaller, or about the same as Mexico's?

_____

Q31) In 2006, was Canada's trade balance with the United States greater, smaller, or about the same as Mexico's?

_____

Q32) Referring to the Attributes of NAFTA_Trading_Statistics table, what was the exact value of Canada's trade balance with the United States in 2006?

_____

Q33) Write the map document's new name and location.

Document _____
(Example: ABC_Region6.mxd)

Location _____
(Example: C:\Student\ABC)

Step 6: Evaluate the effectiveness of NAFTA

Q34) Determine from the graphs the approximate value of exports in 2006 for each pair of countries in the table below. Record the values in the middle column.

Direction of export flow	Value of exports (million $)	Total volume between partners (million $)
United States to Mexico		
Mexico to United States		
United States to Canada		
Canada to United States		
Canada to Mexico		
Mexico to Canada		

Q35) Add the export values together for each pair of countries (for example, exports from the United States to Mexico plus exports from Mexico to the United States). Record the totals in the last column in the table in Q34.

Q36) Rank the trading partners by total volume of trade. Use 1 for the partners trading the most and 3 for the partners trading the least.

United States–Mexico: _____

United States–Canada: _____

Canada–Mexico: _____

Q37) Did NAFTA have a positive (+), negative (–), or neutral (N) effect on trade volume between partner countries?

United States–Mexico: _____

United States–Canada: _____

Canada–Mexico: _____

Q38) Do you think that any one of these three countries benefited more than the other two from NAFTA? If so, which country? Explain your answer.

_____

_____

_____

Q39) What country has a healthier trade balance with Canada: Mexico or the United States?

_____

Q40) On what graph do you find a set of bars that looks like a mirror image of those for the U.S. trade balance with Canada?

_____

Q41) What country had the most dramatic change for the better in trade after NAFTA came into being in 1994?

_____

Q42) Estimate U.S. trade deficits with Canada and Mexico for 2004 and 2006 and record them (in billion dollars) in the table below.

Trading partner	U.S. trade deficit ($)	
	2004	2006
Mexico		
Canada		
Combined		

Q43) Did the U.S. combined trade balance get better or worse between 2004 and 2006?

By how much? _____

Middle school assessment
Module 6, Lesson 2

Share and share alike

1. Using the 1991–2006 trade data for Canada, Mexico, and the United States, determine whether NAFTA has been effective, slightly effective, or ineffective and whether it should be continued or discontinued. You can compare the exports of the three countries, look at their trade balances, and examine other factors that may be relevant. Refer back to the NAFTA Objectives handout to remind yourself of the original intent of NAFTA.

2. Create a layout in ArcMap that graphically displays your conclusion. Use the layout in the Region6_Assessment map document that has already been started for you. You may rearrange and resize the existing layout components and add your own. Your final layout should include the following components:

   - Title
   - Map
   - Map legend
   - Orientation (north arrow or compass rose)
   - At least two graphs
   - Text labels or descriptions
   - Author (your name)
   - Today's date

3. Present your layout to the class as evidence for your conclusion on NAFTA's effectiveness.

M6
L2

High school assessment
Module 6, Lesson 2

Share and share alike

1. Using the 1991–2006 trade data for Canada, Mexico, and the United States, determine whether NAFTA has been effective, slightly effective, or ineffective and whether it should be continued or discontinued. You can compare the exports of the three countries, look at their trade balances, and examine other factors that may be relevant. Consider factors at the global, regional, and local scales. Refer back to the NAFTA Objectives handout to remind yourself of the original intent of NAFTA.

2. Create a layout in ArcMap that graphically displays your conclusion. Use the layout in the Region6_Assessment map document that has already been started for you. You may rearrange and resize the existing layout components and add your own. Your final layout should include the following components:

   - Title
   - Map
   - Map legend
   - Orientation (north arrow or compass rose)
   - At least two graphs
   - Text labels or descriptions
   - Author (your name)
   - Today's date

3. Present your layout to the class as evidence for your conclusion on NAFTA's effectiveness.

4. Compose a paragraph describing how you would change NAFTA to improve or enhance trade for all three countries in the future. Include specific reasons why you believe your changes to NAFTA would be effective.

M6
L2

Module 7, Lesson 1

Water world

A global perspective

- Activity
- Answer sheet
- Assessment
- Assessment table: Data sources

Module 7, Lesson 1

Water world

Imagine that the year is 2100. Scientists have determined that the rapidly warming climate of the earth will cause the ice sheets of Antarctica to break apart and melt at a much faster rate than was predicted a hundred years earlier. You and your GIS investigation team are presented with the challenge of studying the impact this change will have on the planet.

You will explore and compare different maps of Antarctica and investigate specific Antarctic sites to learn more about the continent. You will also use world maps to investigate changes in ocean levels associated with the melting of the Antarctic ice sheets.

Step 1: Open a map document

1. Double-click the ArcMap icon on your computer's desktop.

2. When the ArcMap start-up dialog box appears, click **An existing map** and click OK.

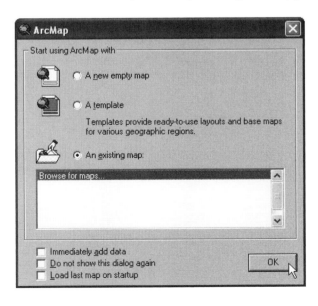

M7
L1

3. Navigate to the module 7 folder (**OurWorld2\Mod7**) and choose **Global7.mxd** (or **Global7**) from the list.

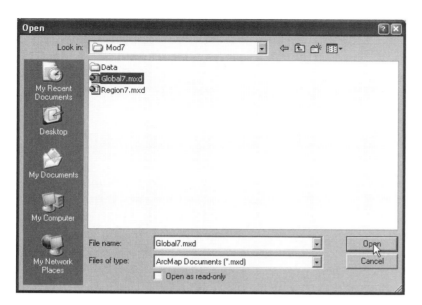

When the map document opens, you see a map with three layers turned on (Latitude & Longitude, Continents, Ocean). The check mark next to the layer name tells you the layer is turned on and visible in the map.

Step 2: Look at Antarctica

1. Use the table of contents to locate the continent of Antarctica.

Take a look at the continent where the melted water will come from. Scientists believe that the first area to melt will be the Western Ice Shelf of Antarctica. The western part of Antarctica is on the left side of the map. It is considerably smaller than the eastern portion. It lies on the west of the Transantarctic Mountain Range and basically covers all the land to the west of the prime meridian. The prime meridian is the line that runs north-south on the map.

Because the map is based on a geographic map projection, which is a flat projection of the spherical earth, some parts of it are skewed (out of shape). This can affect the size of features (such as landforms), their shape, or the distance or direction between them.

> Answers to questions in this activity should be recorded on the answer sheet.

Q1 *Do you think this map gives you a realistic representation of Antarctica? Explain.*

2. At the top of the table of contents, right-click South Pole and click Properties.

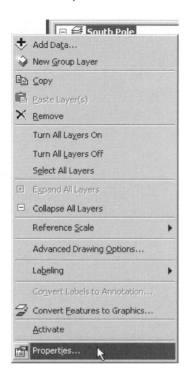

3. Click the Coordinate System tab. The current coordinate system is GCS_WGS_1984 (Geographic Coordinate System, World Geodetic System of 1984).

4. Under "Select a coordinate system," click the plus signs next to the following folders to expand them: Predefined, Projected Coordinate Systems, and World. Scroll down the list of world projections and click Mercator (world).

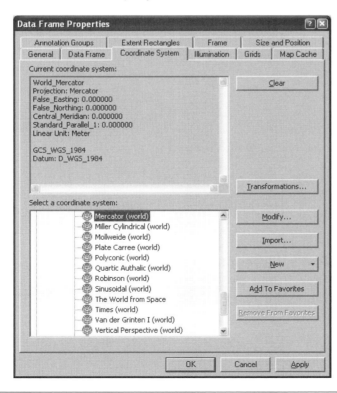

5.   Click OK, then click the Full Extent button to see all the continents.

The map looks rectangular.

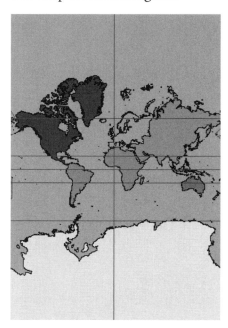

> **Q2** *Does this projection give you a better view of the South Pole region? Why or why not?*

6.   Repeat steps 2–5 to change the map to each of the following projections:
- Aitoff (world)
- Bonne (world)
- Cylindrical equal area (world)
- Equidistant conic (world)
- Mollweide (world)
- Robinson (world)
- Sinusoidal (world)
- Van der Grinten I (world)

> **Q3** *Do any of these projections work well for viewing Antarctica?*

Step 3:   View the South Pole

As you reviewed the various projections, you may have thought that none of them would give you a good perspective of the South Pole, or you may have wanted to flip the map upside down or change its center. The recommended projection is the polar orthographic projection, which centers the map on the South Pole.

1.   Open South Pole Data Frame Properties as you did previously. Make sure that the Coordinate Systems tab is selected.

2. Expand the following folders: Predefined, Projected Coordinate Systems, and Polar. Scroll down the list of polar projections and click South Pole Orthographic.

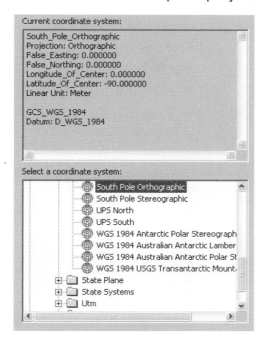

3. Click OK, then click the Full Extent button.

Step 4: Picture Antarctica

1. In the table of contents, click the box next to Antarctic Sites.

2. In the table of contents, right-click Antarctic Sites and click Zoom To Layer to get a closer view of Antarctica.

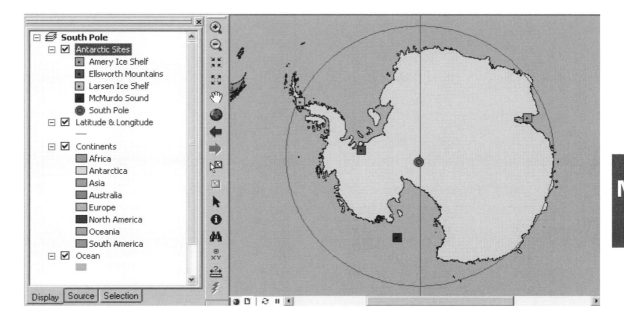

M7
L1

Each of the five sites marked on the map is hyperlinked to an image. As with Internet links, if you click a hyperlink in ArcMap, you are taken to additional information.

 3.   Click the Hyperlink tool. When you move the cursor over the map, it turns into a lightning bolt. Notice that when the Hyperlink tool is selected, a blue dot displays at the center of each Antarctic site, indicating that it has an active hyperlink.

4.   Move the tip of the lightning bolt over the South Pole until the lightning bolt changes to black, then click. A photograph of the South Pole opens in the default image viewer (if the photograph appears cut off, you may need to resize the image or the image viewer).

5.   Read the caption at the bottom of the photograph to learn more about the picture.

6.   Minimize the image viewer.

7.   Find out more about what Antarctica looks like by clicking on the other points on the map. Read the caption at the bottom of each image. You can stretch or maximize the images and the image viewer to see the images and captions at a larger size.

8.   In the table of contents, click the box next to Antarctic Sites to turn it off.

Step 5:   Activate the Water World data frame

Antarctica has two major ice sheets: the western and the eastern. The western sheet is smaller than the eastern and covers Antarctica from the Transantarctic Mountains westward. The eastern sheet is on the opposite side of the mountain range and includes the majority of the continent. Both of these enormous sheets of ice are moving from the continental center toward the ocean. For example, as the western ice sheet moves into the ocean, it forms the Ross and Ronne ice shelves, which float on top of the ocean. It is here that the ice begins to break apart and melt.

You will now examine what might happen to the water levels of the oceans if parts of these ice sheets were to melt.

1.   Click the minus sign next to the South Pole data frame to collapse it in the table of contents.

2.   Click the plus sign next to the Water World data frame to expand its contents.

3.   In the table of contents, right-click Water World and click Activate. A world map in Robinson projection is displayed. The Robinson projection is commonly used for world maps.

You see country outlines in the year 2007. You also see a layer named "20,000 Years Ago." This layer shows an elevation map of the earth as scientists believe it looked 20,000 years ago. At that time, sea level was 400 feet lower than it is today.

4. Stretch the ArcMap window by dragging the lower right corner with your mouse (if the window is not filling the screen already).

5. Zoom and pan the map as needed to answer the following question.

> **Q4** *What significant differences do you see between current landmass outlines and those of 20,000 years ago? List at least three.*

Step 6: Analyze global sea levels that would result if Antarctic ice sheets melted

If the western ice sheet melted, scientists predict that the oceans would rise about 5 meters. If the eastern ice sheet melted, sea level would rise about 50 meters. If all the ice at the South Pole melted, including all the ice shelves and glaciers, sea level would increase by 73 meters.

One by one, you will turn on the layers Today, Plus 5 Meters, Plus 50 Meters, and Antarctic Total Thaw and make observations in the table on your answer sheet. Remember, ArcMap draws the layers starting with the bottom of the table of contents and moves upward. Therefore, a layer that's turned on at the top of the table of contents will "draw over" a layer below it.

> **Q5** *Turn the layers on and off and compare the changes in sea level (zoom in or out if you need to). Record your general observations for each layer in the table on the answer sheet.*

Step 7: View changes in water levels

1. Click the Full Extent button. Turn off all layers except Plus 50 Meters.

2. Click the Add Data button.

3. Navigate to the module 7 LayerFiles folder (**OurWorld2\Mod7\Data\LayerFiles**). Click **Lakes.lyr**, hold the Ctrl key down, and click **Rivers.lyr**. Click Add.

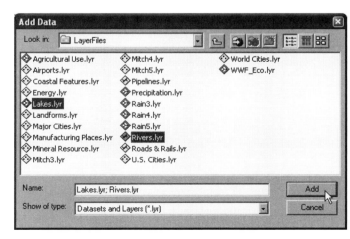

Both layers now appear in the table of contents.

4. Use the Zoom In tool to drag a box around South America.

Q6 *What kinds of changes do you see in the rivers and lakes? Provide a specific example.*

Q7 *With a sea level increase of 50 meters, what kinds of consequences do you foresee for the major river ecosystems of South America? Provide a specific example.*

Q8 *Some inland areas around the globe are below current sea level. One of them is in South America. Hypothesize how these low-lying areas were formed.*

5. Click the Full Extent button to see the entire world on the map.

6. Turn off the Rivers and Lakes layers.

Step 8: View changes in political boundaries

The oceans of the world form the coastlines of many nations. You will now focus on coastal boundaries and how the 50-meter rise would affect political boundaries.

1. Click the Add Data Button.

2. Navigate to the module 7 LayerFiles folder (**OurWorld2\Mod7\Data\LayerFiles**) and double-click **Major Cities.lyr**.

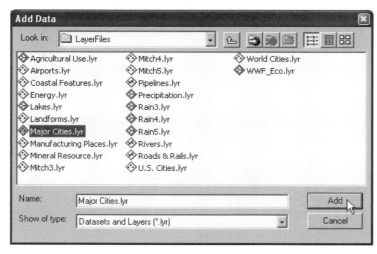

3. Click the Zoom In tool. Zoom to focus on the Southwest Asia

4. Turn on the Major Cities layer.

The dark blue dots represent cities. Note how some of them are now in the water or on water's edge.

5. Turn on Country Outlines so you can view current country boundaries. Take note of significant changes in the amounts of land remaining in different countries.

(Q9) *Predict possible consequences of the 50-meter rise in sea level to the populations living in the Southwest Asia (political disputes, trade and economic issues, transportation problems, etc.). Record those consequences in the first row of the table on the answer sheet.*

6. Click the Full Extent button. Repeat the process of zooming and identifying potential consequences of the rising sea level for the other major regions of the world (use the Zoom, Pan, and Identify tools as needed).

(Q10) *Record your predictions in the table in Q9.*

7. Click the Full Extent button.

(Q11) *List other possible layers of data you might want to analyze to study the impact of rising sea levels.*

8. Ask your teacher for instructions on where to save this map document and how to rename it.

9. If you are not going to save the map document, exit ArcMap by choosing Exit from the File menu. When asked if you want to save changes to Global7.mxd (or Global7), click No.

In this activity, you used ArcMap to investigate the continent of Antarctica. You also explored the potential effect that the thawing of the Antarctic ice sheets would have on the global environment.

Name_____ Date_____

Module 7, Lesson 1

Water world

Step 2: Look at Antarctica

Q1)   Do you think this map gives you a realistic representation of Antarctica? Explain.

_____

_____

_____

_____

_____

Q2)   Does this projection give you a better view of the South Pole region? Why or why not?

_____

Q3)   Do any of these projections work well for viewing Antarctica? _____

Step 5: Activate the Water World data frame

Q4)   What significant differences do you see between current landmass outlines and those of 20,000 years ago? List at least three.

_____

_____

_____

_____

_____

Step 6: Analyze global sea levels that would result if Antarctic ice sheets melted

Q5) Record your general observations for each layer in the table below.

Sea level	Observations
Today	
Plus 5 meters	
Plus 50 meters	
Total thaw (plus 73 meters)	

Step 7: View changes in water levels

Q6) What kinds of changes do you see in the rivers and lakes? Provide a specific example.

_____

_____

_____

_____

_____

Q7) With a sea level increase of 50 meters, what kinds of consequences do you foresee for the major river ecosystems of South America? Provide a specific example.

_____

_____

_____

_____

_____

Q8) Some inland areas around the globe are below sea level. One of them is in South America. Hypothesize how these low-lying areas were formed.

_____

_____

_____

_____

_____

_____

Step 8: View changes in political boundaries

Q9) Predict possible consequences of the 50-meter rise in sea level to the populations living in the Southwest Asia (political disputes, trade and economic issues, transportation problems, etc.). Record those consequences in the first row of the table below.

Region	Countries/areas affected	Possible consequences
Southwest Asia		
Asia		
Europe		
Africa		
Oceania		
North America		
Latin America		

Q10) Record your predictions in the table above.

Q11) List other possible layers of data you might want to analyze to study the impact of rising sea levels.

_____

_____

_____

_____

_____

_____

Middle school assessment
Module 7, Lesson 1

Water world

You and your teammates have been selected to be part of an elite team of GIS experts who will determine the fate of a major city. Over the next 50 years, a rise in sea level of up to 50 meters will affect:

San Francisco, USA	Miami, USA
London, England	Calcutta, India
Tokyo, Japan	Houston, USA

Select a city and develop an action plan for relocating the city and its resources. The plan must take into account the following factors:

Major roads	Ocean ports
Railroads	Utilities
Airports	Relocation of people

The available data sources are listed in the Assessment table "Data Sources." You will add this data to the Water World data frame in the Global7 map document you used in the activity.

Your action plan must include each of the following:
- A time line describing the various phases of your plan. For example, one five-year phase might include relocating people, while another may deal with relocating specific businesses.
- A map displaying proposed changes. This could be a series of maps generated in ArcMap or on paper.
- Data supporting your suggested changes. This data will come from the activity and from the sources listed in the Assessment table "Data Sources" and can be displayed in maps, charts, or tables.
- A written report explaining your plan.

M7
L1

High school assessment
Module 7, Lesson 1

Water world

You and your teammates have been selected to be part of an elite team of GIS experts who will determine the fate of a major city. Over the next 50 years, a rise in sea level of up to 50 meters will affect:

San Francisco, USA	Houston, USA
London, England	Odessa, Ukraine
Tokyo, Japan	Rome, Italy
Miami, USA	Sydney, Australia
Calcutta, India	Buenos Aires, Argentina

Select a city and develop an action plan that relocates the city, relocates the roles of the city to another city, adapts the city to its new environment, or develops another strategy. The plan must take into account the following factors:

Major roads	Utilities
Railroads	Relocation of people
Airports	Economics and trade relations
Ocean ports and shipping lanes	Agriculture and manufacturing

The available data sources are listed in the Assessment table "Data Sources." You will add this data to the Water World data frame in the Global7 map document you used in the activity. You may need to consult an atlas or the Internet to research some factors.

Your action plan must include each of the following:

- A time line describing the various phases of your plan. For example, one five-year phase might include relocating people, while another may deal with relocating specific businesses.
- A map displaying proposed changes. This could be a series of maps generated in ArcMap or on paper.
- Data supporting your suggested changes. This data will come from the activity and from the sources listed in the Assessment table "Data Sources." and can be displayed in maps, graphs, or tables.
- A written report explaining your plan.

M7
L1

Assessment table: Data Sources

The layers below can be found in the module 7 LayerFiles folder
(**OurWorld2\Mod7\Data\LayerFiles**).

Layer	Data
Roads & Rail	Lines that represent roads and railroads
World Cities	Points that represent major world cities and indicate shipping ports
U.S. Cities	Points that represent U.S. cities
Airports	Points that represent airports
Energy	Points that represent major power plants and indicate the types of energy (atomic, thermal, etc.)
Pipelines	Lines that represent major oil and gas pipelines
Manufacturing Places	Points that represent major manufacturing places
Mineral Resources	Points that represent mineral mining sites
Rivers	Lines that represent major rivers
Lakes	Polygons that represent major lakes

M7
L1

Module 7, Lesson 2

In the eye of the storm

A regional investigation of Central America

- Activity
- Answer sheet
- Assessment

Module 7, Lesson 2

In the eye of the storm

October 21, 1998

A tropical storm is brewing in the Atlantic Ocean. It began as a tropical wave a few weeks earlier, off the coast of western Africa. Today it is causing some rain and thunderstorms over the Caribbean. Later, the barometric pressure of the system will continue to drop and it will soon be identified as a tropical depression—the beginning of a hurricane. By the time Hurricane Mitch leaves the Central America region, more than 11,000 people will have died and between 11,000 and 18,000 will be missing.

Step 1: Open a map document

1. Double-click the ArcMap icon on your computer's desktop.

2. When the ArcMap start-up dialog box appears, click **An existing map** and click OK.

3. Navigate to the module 7 folder (**OurWorld2\Mod7**) and choose **Region7.mxd** (or **Region7**) from the list.

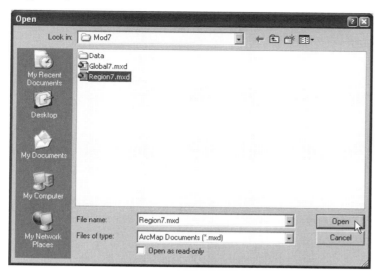

4. Click Open. The map document opens and you see a world map.

5. Scroll down in the table of contents.

   Check marks next to Lakes, Central America, Continents, and Ocean tell you that these layers are displayed on the map.

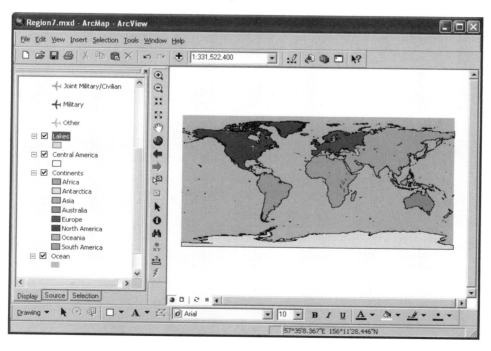

Step 2: Identify the capital cities of Central America

1. Right-click on the Central America layer name and click Zoom to Layer. The map centers on Central America.

Before looking at the effects of Hurricane Mitch, you will collect data on Central America prior to Mitch.

2. Scroll to the top of the table of contents and turn on the Capitals layer by clicking the box next to the layer name.

You can find the names of the capitals by using the Identify tool or by labeling features. Labeling features is a quick way to get information about a group of features.

3. In the table of contents, right-click Capitals and click Properties. Click the Labels tab.

4. At the top of the Labels tab, click the small white box next to "Label features in this layer." Notice that NAME is already chosen as the field to use for labeling.

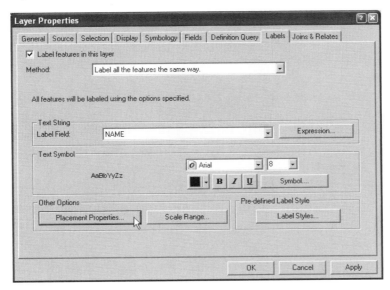

5. Near the bottom of the dialog box under Other Options, click the Placement Properties button. Click the Conflict Detection tab. At the bottom of the tab, click the small white box next to "Place overlapping labels."

Placing overlapping labels will ensure that all of the capitals will be labeled.

6. Click OK on the Placement Properties dialog box and the Layer Properties dialog box. The name of each capital city is displayed on the map.

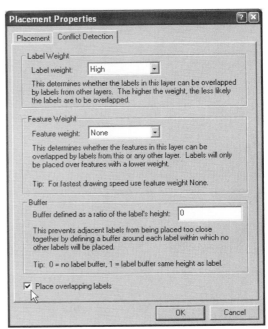

M7
L2

Q1 *Record the capitals of Central American countries in the table on the answer sheet (if you don't know the name of a country, use MapTips to identify it). The information for the country of Belize has been completed for you as an example of the data you will need to find.*

7. Right-click Capitals in the table of contents and click Label Features. The labels disappear from the map.

Step 3: Investigate Central America prior to Hurricane Mitch

1. Turn on the following layers and analyze the geographic distribution patterns that you see: Populated Places, Roads, Railroads, and Airports.

Q2 *Fill in the Populated places Distribution column and the three Transportation columns in the table in Q1.*

Remember, layers that are near the top of the table of contents will cover up layers that are listed lower. Turn on each layer individually to see it clearly. Hover over a country to see the country name.

 2. Click the Add Data button.

3. Navigate to the module 7 LayerFiles folder (**OurWorld2\Mod7\Data\LayerFiles**).

4. Hold down the Ctrl key and click **Agricultural Use.lyr**, **Precipitation.lyr**, and **Landforms.lyr**.

5. Click Add. The added polygon layers Agricultural Use and Precipitation appear below the Airports layer, and the added point layer Landforms appears at the top of the table of contents. The three layers are turned off.

6. Turn on Precipitation. Turn off Railroads, Populated Places, Roads, and Airports.

    The annual precipitation data appears on the map, but you cannot determine the average precipitation for each country. You will move the Central America layer and change its legend so you can view each country's average precipitation data.

7. Click and drag the Central America layer above Agricultural Use in the table of contents. The Central America layer now covers up the Precipitation layer.

8. Click on the yellow symbol for the Central America layer to open the Symbol Selector.

9. Click the Hollow symbol.

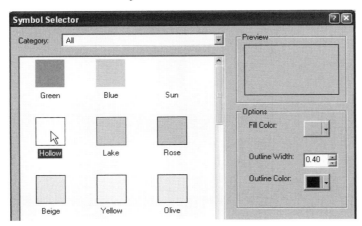

10. In the Options panel on the right, increase the Outline Width to 1.5.

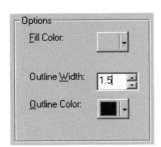

11. Click OK. The map now displays Central American countries as outlines.

Q3 *Analyze the annual precipitation for each country and fill in the Precipitation column in the table in Q1.*

12. Turn off Precipitation and turn on Agricultural Use.

Q4 *Fill in the last column in the table in Q1.*

Q5 *Which country has the largest proportion of its area devoted to crops?*

13. Turn off the Agricultural Use layer and turn on the Landforms layer.

Q6 *Which country is the most mountainous?*

14. Turn off Landforms and turn on Roads.

Q7 *Which country has the largest proportion of its territory covered by roads?*

15. Click the minus sign to the left of Snapshot of Central America to collapse the data frame.

M7
L2

Q8 *If you are stopping here, save the map document under a new name, record the new name and location, and exit ArcMap.*

Step 4: Track Hurricane Mitch

October 24–26, 1998

In a span of less than two days, Tropical Storm Mitch develops into a category 5 hurricane with winds in excess of 155 knots (about 180 miles per hour). Category 5 is the deadliest rating on the Saffir-Simpson Hurricane Potential Damage Scale. Barometric pressure dropped to 905 millibars, tying with Hurricane Camille (1969) for the lowest pressure ever observed in the Atlantic basin.

1. If you exited at the end of the last section, reopen ArcMap and navigate to the map document you saved.

2. In the table of contents, right-click the Hurricane Mitch data frame and click Activate. Click the plus sign to expand the contents.

The layers Latitude & Longitude, Central America, Continents, and Ocean are turned on.

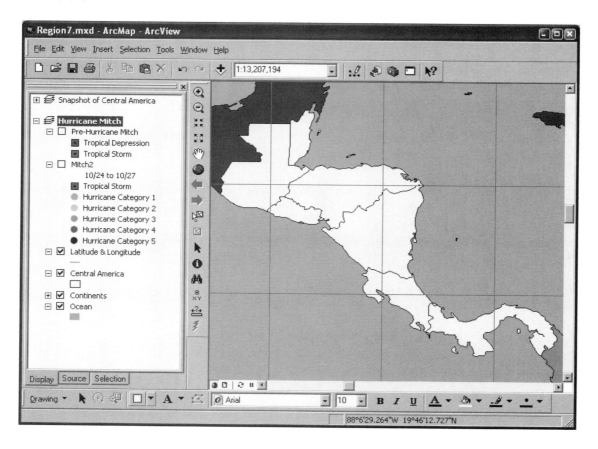

3. Turn on Pre-Hurricane Mitch and Mitch2.

Both of these layers (and similarly named layers to follow) show the locations of the center, or eye, of the storm at different time points. For locations where Mitch was a hurricane, the legend indicates Mitch's category. Placing your mouse pointer over a location displays the MapTip for that location.

4. Click the Identify tool. Click the most southeastern dark square representing Tropical Storm Mitch.

An Identify window displays latitude and longitude (LAT and LON), time, wind velocity in miles per hour, pressure (PR) in millibars, and status on the Saffir-Simpson scale. The time is given in the format month/day/hour (24-hour clock); the Z stands for Zulu time, which is the time at 0° longitude, used as a standard reference for anywhere on the globe.

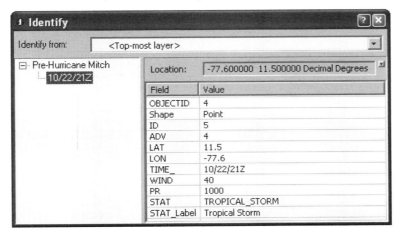

Q9 *When was Tropical Storm Mitch at this location?*

Q10 *What was Mitch's wind speed at this location?*

5. Click the location of category 1 Hurricane Mitch on the map. The Identify window updates.

Q11 *What are the latitude and longitude coordinates for Hurricane Mitch at this location?*

Q12 *When was Hurricane Mitch at this location?*

Q13 *What was Mitch's wind speed at this location?*

6. Click the last mapped location of category 5 Hurricane Mitch.

Q14 *When was Hurricane Mitch at this location?*

Q15 *What was Mitch's wind speed at this location?*

M7
L2

Now you will determine how much time it took Mitch to develop from a tropical storm to a category 5 hurricane.

7. Close the Identify window.

8. In the table of contents, right-click Mitch2 and click Open Attribute Table. The Attributes of Mitch2 table opens.

9. Click the TIME_ column heading to select it. The heading depresses like a button, and the column turns light blue.

10. Right-click the TIME_ column heading and click Sort Ascending.

11. Click the small gray box to the left of the first record to select it. The first record turns light blue to show that it is selected. Notice that this record represents Hurricane Mitch when it was a tropical storm.

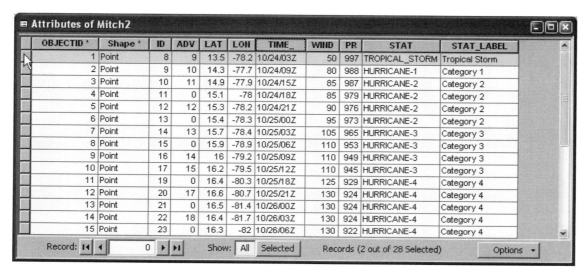

OBJECTID *	Shape *	ID	ADV	LAT	LON	TIME_	WIND	PR	STAT	STAT_LABEL
1	Point	8	9	13.5	-78.2	10/24/03Z	50	997	TROPICAL_STORM	Tropical Storm
2	Point	9	10	14.3	-77.7	10/24/09Z	80	988	HURRICANE-1	Category 1
3	Point	10	11	14.9	-77.9	10/24/15Z	85	987	HURRICANE-2	Category 2
4	Point	11	0	15.1	-78	10/24/18Z	85	979	HURRICANE-2	Category 2
5	Point	12	12	15.3	-78.2	10/24/21Z	90	976	HURRICANE-2	Category 2
6	Point	13	0	15.4	-78.3	10/25/00Z	95	973	HURRICANE-2	Category 2
7	Point	14	13	15.7	-78.4	10/25/03Z	105	965	HURRICANE-3	Category 3
8	Point	15	0	15.9	-78.9	10/25/06Z	110	953	HURRICANE-3	Category 3
9	Point	16	14	16	-79.2	10/25/09Z	110	949	HURRICANE-3	Category 3
10	Point	17	15	16.2	-79.5	10/25/12Z	110	945	HURRICANE-3	Category 3
11	Point	19	0	16.4	-80.3	10/25/18Z	125	929	HURRICANE-4	Category 4
12	Point	20	17	16.6	-80.7	10/25/21Z	130	924	HURRICANE-4	Category 4
13	Point	21	0	16.5	-81.4	10/26/00Z	130	924	HURRICANE-4	Category 4
14	Point	22	18	16.4	-81.7	10/26/03Z	130	924	HURRICANE-4	Category 4
15	Point	23	0	16.3	-82	10/26/06Z	130	922	HURRICANE-4	Category 4

Record: 0   Show: All Selected   Records (2 out of 28 Selected)   Options ▾

12. Scroll down and locate the record representing the first time Hurricane Mitch became a category 5 hurricane (look in the STAT column). Hold the Ctrl key down and click the small gray box for the Hurricane-5 record. Both records are highlighted blue in the table and on the map.

Remember, you must click the small gray box to the left of a row in the table in order to select the entire row.

13. Click the Selected button at the bottom of the attribute table. Now you see only the two selected records, and they are easier to compare.

Q16 *How long did it take for Tropical Storm Mitch to become a category 5 hurricane? On the answer sheet, write down the time for each event and determine the time difference. Remember, the time is given in the format month/day/hour (24-hour clock).*

14. Click the All button at the bottom of the attribute table to see all the records again.

Q17 *What was Hurricane Mitch's maximum wind speed during this period?*

15. Close the Attributes of Mitch2 table.

Step 5: Measure the size of the storm

The National Oceanic and Atmospheric Administration (NOAA), in partnership with the National Aeronautics and Space Administration (NASA), used special storm-tracking satellites to take several high-resolution photographs of Mitch from space. You will view these images and measure the massive size of this storm.

 1. Click the Add Data button and navigate to the module 7 Images folder (**OurWorld2\ Mod7\Data\Images**).

2. Click mitch2sat.tif and click Add.

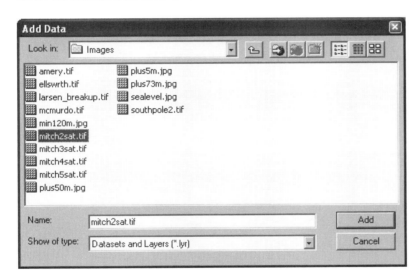

The satellite image now sits underneath the map, and you cannot see it. In order to see both the satellite image and the storm track, you will rearrange the layers in the table of contents.

3. Click and drag the mitch2sat.tif layer so that it's just below Mitch2 in the table of contents.

Now you can see the highlighted locations of Tropical Storm Mitch and category 5 Hurricane Mitch. The category 5 point is almost directly over the eye of the storm on the satellite image. The eye is the center of the cloud mass and looks like a doughnut hole.

4. Turn off Mitch2 so you have a better view of the eye of Hurricane Mitch.

 5. Use the Zoom In and Pan tools so the satellite image fills the map display. Do not zoom in too close or it will be difficult to view the image.

M7
L2

6. Click the Measure tool. The measure dialog box opens, and your cursor turns into a right-angle ruler with crosshairs

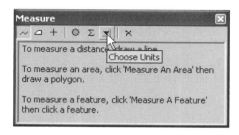

7. Click on the Choose Units button, Distance, and Miles.

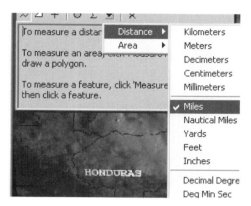

8. Click the left edge of the eye once and move the cursor directly across the eye. Double-click when your cursor is at the right edge of the eye (if you accidentally clicked the wrong spot, you can double-click to end the line and simply start over).

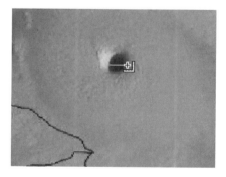

The length of the segment in miles appears in the Measure dialog box.

Q18 *What is the diameter of the eye of Hurricane Mitch?*

9. Measure the total diameter of the storm at its widest point and the distance between the eye and the coastline of Honduras (use the Zoom and Pan tools as needed).

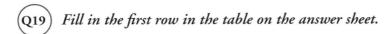 **Q19** *Fill in the first row in the table on the answer sheet.*

10. Close the Measure dialog box. Turn off mitch2sat.tif.

   Now that you have recorded data for mitch2sat.tif, you will follow the same procedure for adding satellite images of Mitch at later time points and measuring the storm as it moved inland.

11. Click the Add Data button. Navigate to the module 7 images folder (**OurWorld2\ Mod7\Data\Images**).

12. Hold down the Ctrl key and add the following images: **mitch3sat.tif**, **mitch4sat.tif**, and **mitch5sat.tif**.

13. Turn off all three images. Click the minus sign in front of each image to collapse the legend.

   You will now add layers showing Mitch's track as it made landfall and moved inland.

14. Click the Add Data button again. Click the Up One Level button and navigate to the module 7 LayerFiles folder (**OurWorld2\Mod7\Data\LayerFiles**).

15. Hold down the Ctrl key and add the following layer files: **Mitch3.lyr**, **Mitch4.lyr**, and **Mitch5.lyr**.

   With six new layers added, it's important to organize your table of contents so you can view the layers easily.

16. In the table of contents, click mitch3sat.tif to unselect the other layers. Click it again and drag it directly below Mitch3. Do this for the Mitch 4 satellite image as well.

```
☐ ☐ Mitch3
         10/28 to 10/29
      ■ Tropical Storm
      ● Hurricane Category 1
      ● Hurricane Category 2
      ● Hurricane Category 3
      ● Hurricane Category 4
      ● Hurricane Category 5
 ☐ ☐ mitch3sat.tif
☐ ☐ Mitch4
         10/30
      ■ Tropical Depression
      ■ Tropical Storm
 ☐ ☐ mitch4sat.tif
☐ ☐ Mitch5
         10/31
      ■ Tropical Depression
      ■ Tropical Storm
 ☐ ☐ mitch5sat.tif
```

**M7
L2**

Q20 *Fill in the rest of the table in Q19. Turn layers on and off as needed and use the Measure, Zoom, and Pan tools.*

Step 6: Analyze rainfall from Hurricane Mitch

Once Hurricane Mitch made landfall, the winds weakened to the point where it was downgraded to a tropical storm. Nonetheless, Mitch still had not done its worst damage. In the days that followed, Mitch poured more than 30 inches of rain on the region. You will now take a closer look at the precipitation that fell on October 30 and 31, 1998.

1. Turn off all layers except mitch3sat.tif, Central America, Continents, and Ocean.

 2. Click the Add Data button. Navigate to the module 7 LayerFiles folder (**OurWorld2\ Mod7\Data\LayerFiles**). Add the following layer files to your map: **Rain3.lyr, Rain4. lyr**, and **Rain5.lyr**.

3. Turn on Rain3. The rain pattern is overlaid on top of mitch3sat.tif.

Q21 *What pattern do you notice in the amount of rainfall?*

Q22 *Is this a pattern you expected to find? Why or why not?*

4. In the table of contents, select only Rain4 and drag it above the corresponding satellite image, mitch4sat.tif. Do this for the other rain layers.

☐ Mitch4
    10/30
    ■ Tropical Depression
    ■ Tropical Storm
☐ Rain4
    Rainfall in inches
    ☐ 0-6
    ■ 7-11
    ☐ 12-17
    ☐ 18-23
    ☐ 24-29
    ■ 30+
☐ mitch4sat.tif
☐ Mitch5
    10/31
    ■ Tropical Depression
    ■ Tropical Storm
☐ Rain5
    Rainfall in inches
    ☐ 0-6
    ■ 7-11

5. Turn rain and satellite layers on and off as needed to answer the following questions.

Q23 *What is the highest range of rainfall in the Rain4 layer?*

Q24 *Which country received the majority of this heavy rain?*

Q25 *Describe the difference between the rainfall patterns on October 30 (Rain4 layer) and October 31 (Rain5 layer).*

Q26 *What kind of damage do you expect to find with this type of storm? What aspects of the region will be most affected? Use the table in Q1 as a resource.*

Q27 *Ask your teacher for instructions on where to save this map document and how to rename it. Record the new name of the document and its location.*

6. Choose Exit from the File menu.

In this activity, you used ArcMap to analyze a large region of Central America and to track Hurricane Mitch as it made landfall. You probably have many questions as to the extent of the damage Mitch caused. In the assessment you will take on the role of emergency management personnel. Your job will be to identify areas where danger from the storm is high and to develop an emergency action plan for one of the affected countries. You will use the data from this lesson.

M7
L2

Name_____ Date_____

Module 7, Lesson 2

In the eye of the storm

Step 2: Identify the capital cities of Central America

Q1)   Record the capitals of Central American countries in the table below.

Central America Prior to Hurricane Mitch

Country	Populated places		Transportation			Precipitation	Agricultural use
	Capital	Distribution	Roads	Railroads	Airports		
Belize	Belmopan	Throughout the country, but concentrated around the capital	Sparse network	None	One near the coast	Primarily 1,401–2,800 mm	Primarily forest, with some irrigated land and little cropland
Guatemala							
Honduras							
El Salvador							
Nicaragua							
Costa Rica							
Panama							

Step 3: Investigate Central America prior to Hurricane Mitch

Q2) Fill in the Populated places Distribution column and the three Transportation columns in the table in Q1.

Q3) Analyze the annual precipitation for each country and fill in the Precipitation column in the table in Q1.

Q4) Fill in the last column in the table in Q1.

Q5) Which country has the largest proportion of its area devoted to crops? _____

Q6) Which country is the most mountainous? _____

Q7) Which country has the largest proportion of its territory covered by roads? _____

Q8) If you are stopping here, save the map document under a new name, record the new name and location, and exit ArcMap.

Document _____
(Example: ABC_Region7.mxd)

Location _____
(Example: C:\Student\ABC)

Step 4: Track Hurricane Mitch

Q9) When was Tropical Storm Mitch at this location? _____

Q10) What was Mitch's wind speed at this location? _____

Q11) What are the latitude and longitude coordinates for Hurricane Mitch at this location?

_____

Q12) When was Hurricane Mitch at this location? _____

Q13) What was Mitch's wind speed at this location? _____

Q14) When was Hurricane Mitch at this location? _____

Q15) What was Mitch's wind speed at this location? _____

Q16) How long did it take for Tropical Storm Mitch to become a category 5 hurricane? Write down the time for each event and determine the difference.

Hurricane–5 time point: _____

Tropical_Storm time point: _____

Time difference: _____

Q17) What was Hurricane Mitch's maximum wind speed during this period? _____

Step 5: Measure the size of the storm

Q18) What is the diameter of the eye of Hurricane Mitch? _____

Q19) Fill in the first row in the table below.

Image	Distance (miles)			Change from previous image
	Diameter of the eye	Diameter of the storm	Between the eye and the coastline of Honduras	
mitch2sat.tif				-------------
mitch3sat.tif				
mitch4sat.tif				
mitch5sat.tif				

Q20) Fill in the rest of the table above.

Step 6: Analyze rainfall from Hurricane Mitch

Q21) What pattern do you notice in the amount of rainfall?

_____

Q22) Is this a pattern you expected to find? Why or why not?

_____

Q23) What is the highest range of rainfall in the Rain4 layer?

_____

Q24) Which country received the majority of this heavy rain?

_____

Q25) Describe the difference between the rainfall patterns on October 30 (Rain4 layer) and October 31 (Rain5 layer).

_____

_____

_____

Q26) What kind of damage do you expect to find with this type of storm? What aspects of the region will be most affected? Use the table in Q1 as a resource.

_____

_____

Q27) Record the new name of the document and its location.

Document _____
(Example: ABC_Region7.mxd)

Location _____
(Example: C:\Student\ABC)

Middle school assessment: In the eye of the storm
Module 7, Lesson 2

Volcano and landslide hazard team

Country _____

Your team is in charge of dealing with the potential hazards of landslides and debris flows from volcanoes in the region devastated by the intense rainfall of Hurricane Mitch. Investigate the country's volcanoes and the amounts of rainfall they received from the storm. Compare this rainfall to the average precipitation for that area. To see the locations of utility lines, add the ca_utility data to the map. It is located in the Central America geodatabase (**Mod7\Data\CentralAmerica.gdb**).

Predict which towns are in the most danger from landslides and debris flows and create an emergency action plan for these towns that addresses:
- Evacuation
- Ways to reroute power
- Provision of medical and humanitarian aid
- Alternatives to damaged transportation networks

Your report should also identify agricultural areas that might be damaged (if any).

**M7
L2**

Middle school assessment: In the eye of the storm
Module 7, Lesson 2

Flood hazard team

Country _____

Your team is in charge of dealing with potential damage from rising floodwaters caused by Hurricane Mitch. You will need to look at the rainfall layers from Hurricane Mitch and typical precipitation patterns for the region. You will also need to add data on drainage features (ca_drain) and utility lines (ca_utility) to the map document. These are in the Central America geodatabase (**Mod7\Data\ CentralAmerica.gdb**).

Predict which towns are in the most danger from flooding and create an emergency action plan for these towns that addresses:
- Evacuation
- Ways to reroute power
- Provision of medical and humanitarian aid
- Alternatives to flooded transportation networks

Your report should also identify agricultural areas that might be damaged (if any).

**M7
L2**

High school assessment: In the eye of the storm
Module 7, Lesson 2

Volcano and landslide hazard team

Country _____

Your team is in charge of dealing with the potential hazards of landslides and debris flows from volcanoes in the region devastated by the intense rainfall of Hurricane Mitch. Investigate the country's volcanoes and the amounts of rainfall they received from the storm. Compare this rainfall to the average precipitation for that area. To see the locations of utility lines, add the ca_utility data to the map. It is located in the CentralAmerica geodatabase (**Mod7\Data\CentralAmerica.gdb**).

Predict which towns are in the most danger from landslides and debris flows and create an emergency action plan for these towns that addresses:
- Evacuation
- Ways to reroute power
- Provision of medical and humanitarian aid to the affected areas
- Alternatives to damaged transportation networks

Your report should also identify agricultural areas that might be damaged (if any) and discuss how this would hinder the economy of the country.

**M7
L2**

High school assessment: In the eye of the storm
Module 7, Lesson 2

Flood hazard team

Country _____

Your team is in charge of dealing with the potential damage from rising floodwaters caused by Hurricane Mitch. You will need to look at the rainfall layers from Hurricane Mitch and typical precipitation patterns for the region. You will also need to add data on drainage features (ca_drain) and utility lines (ca_utility) to the map document. These are in the Central America geodatabase (**Mod7\Data\ CentralAmerica.gdb**).

Predict which towns are in the most danger from flooding and create an emergency action plan for these towns that addresses:
- Evacuation
- Ways to reroute power
- Provision of medical and humanitarian aid
- Alternatives to flooded transportation networks

Your report should also identify agricultural areas that might be damaged (if any) and discuss how this would hinder the economy of the country.

**M7
L2**

ArcMap toolbar reference

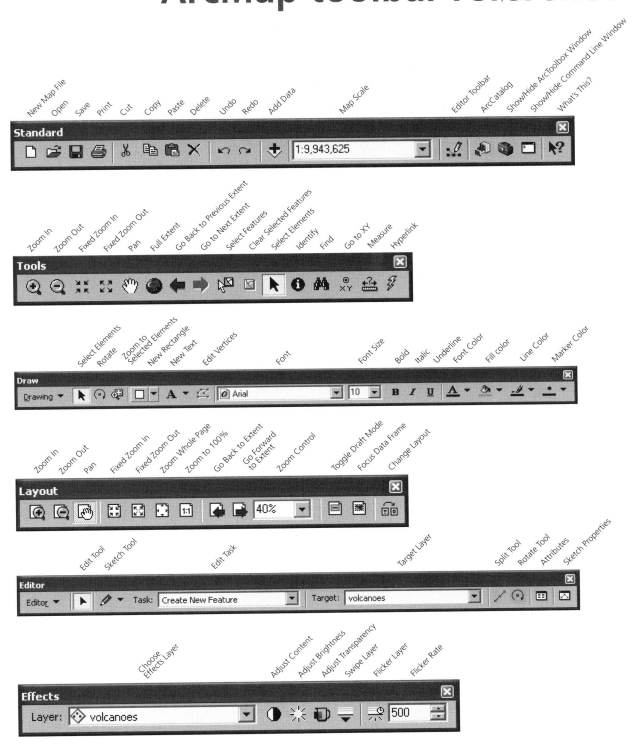

ArcMap zoom and pan tools

ArcMap gives you a number of ways to move around the map display. The better you know these tools, the more quickly you will be able to investigate different areas and features. Here are a few of them:

Tools for zooming in or out from the center of the map

Fixed Zoom In. Click the button to zoom in a fixed amount toward the center of the map. Click the button again to zoom in more.

Fixed Zoom Out. Click the button to zoom out a fixed amount away from the center of the map. Click the button again to zoom out more.

Tools for zooming in or out from anyplace on the map

Zoom In. Click the button, then click a spot on the map or drag a box around an area to zoom in on it. When the map redraws, the point or area you selected will appear in the center of the display.

Zoom Out. Click the button, then click a spot on the map or drag a box around an area to zoom out from it. When the map redraws, the point or area you selected will appear in the center of the display.

Tools for jumping to a specific map display

Full Extent. Click the Full Extent button to zoom to the entire map.

Go Back To Previous Extent. Click the button one or more times to go backward through earlier map displays you were browsing.

Go To Next Extent. Click the button one or more times to go forward again through the series of map displays you were browsing.

Tools for panning the map

Pan. Click the button. Click a spot on the map, hold down the mouse button, and drag it to a new location.

Scrollbars. Drag the scrollbars below and to the right of the map to pan the map side to side or up and down.

Zooming and panning with your mouse

If your mouse has a center wheel (a wheel between the left and right mouse buttons) you can use the wheel to zoom and pan the map display.

Zoom. Click a spot on the map. Roll the mouse wheel forward and backward to zoom in or out.

Pan. Move your mouse pointer over the map. Click with the mouse wheel and hold it down. Drag the mouse to pan the map.

Making quality maps

Think about the items below as you finish your maps or layouts. They will help you make sure your maps and layouts are the best they can be.

Map composition

Do your maps have the following elements?
- Title (addresses the major theme in your analysis)
- Legend
- Scale
- Compass rose
- Author (your name)
- Date map was created

Classification

Did you make reasonable choices for the classifications of the different layers on your maps? Is the symbology appropriate for the various layers?
- For quantitative data, is there a logical progression from low to high values and are they clearly labeled?
- For qualitative data, did you make sure not to imply any ranking in your legend?

Scale and projection

- Is the map scale appropriate for your problem?
- Have you used an appropriate map projection?

Implied analysis

- Did you correctly interpret the color, pattern, and shape of your symbologies?
- Have you included text to inform the reader of the map's intended use?

Design and aesthetics

- Are your maps visually balanced and attractive?
- Can you distinguish the various symbols for different layers in your maps?

Effectiveness of map

- How well do the map components communicate the story of your map?
- Do the map components take into account the interests and expertise of the intended audience?
- Are the map components of appropriate size?

GIS terms

ArcGIS

Computer software for implementing a geographic information system (GIS).

ArcView

Desktop GIS software that includes ArcMap for displaying and interacting with maps and layouts and ArcCatalog for previewing data and metadata.

attribute

A piece of information that describes a geographic feature on a GIS map. The attributes of an earthquake might include the date it occurred and its latitude, longitude, depth, and magnitude.

attribute table

A table that contains all of the attributes for like features on a GIS map, arranged so that each row represents one feature and each column represents one feature attribute. In a GIS, attribute values in an attribute table can be used to find, query, and symbolize features.

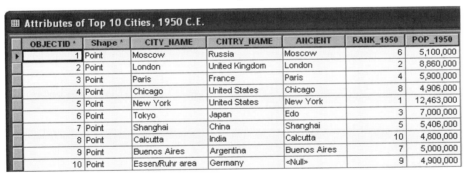

OBJECTID *	Shape *	CITY_NAME	CNTRY_NAME	ANCIENT	RANK_1950	POP_1950
1	Point	Moscow	Russia	Moscow	6	5,100,000
2	Point	London	United Kingdom	London	2	8,860,000
3	Point	Paris	France	Paris	4	5,900,000
4	Point	Chicago	United States	Chicago	8	4,906,000
5	Point	New York	United States	New York	1	12,463,000
6	Point	Tokyo	Japan	Edo	3	7,000,000
7	Point	Shanghai	China	Shanghai	5	5,406,000
8	Point	Calcutta	India	Calcutta	10	4,800,000
9	Point	Buenos Aires	Argentina	Buenos Aires	7	5,000,000
10	Point	Essen/Ruhr area	Germany	<Null>	9	4,900,000

Attributes of Top 10 Cities, 1950 C.E.

The attribute table for the Top 10 Cities, 1950 C.E. layer includes attributes for each of the ten cities listed.

axis

The vertical (y-axis) or horizontal (x-axis) lines in a graph on which measurements can be illustrated and coordinated with each other. Each axis in a GIS graph can be made visible or invisible and labeled.

bookmark

In ArcMap, a shortcut you can create to save a particular geographic extent on a map so you can return to it later. Also known as a spatial bookmark.

color selector

The window that allows you to change the color of geographic features and text on your GIS map.

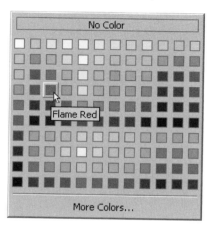

comma-delimited values file (.csv)

A data table in text form where the values are separated by commas. This is a popular format for transferring data from one program to another, for example between spreadsheet programs and ArcMap. These programs use the commas to determine where a new piece of data starts and stops.

coordinate system

A system of intersecting lines that is used to locate features on surfaces such as the earth's surface or a map. In ArcMap, each feature class (layer) of data has a coordinate system that tells ArcMap where on the map to draw the features. A feature class may also have a map projection. See also feature class and map projection.

data

Any collection of related facts, from raw numbers and measurements to analyzed and organized sets of information.

data folder

A folder on the hard drive of your computer or your network's computer that is available for storage of GIS data and map documents that you create.

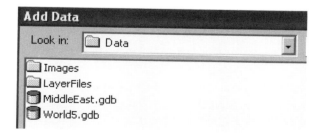

data frame

A map element that defines a geographic extent, a page extent, a coordinate system, and other display properties for one or more layers in ArcMap. In data view, only one data frame is displayed at a time; in layout view, all of a map's data frames are displayed at the same time.

data frame, active

In ArcMap, the active data frame is the target for many tools and commands. In data view, the active data frame is bold in the table of contents and visible in the display area. In layout view, the active data frame has a dashed line around it to show it is the active one.

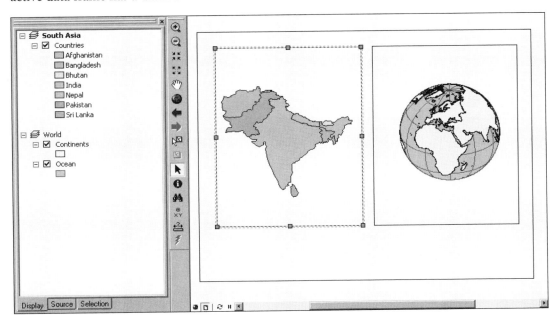

This map document, shown in layout view, has two data frames: South Asia and The World. South Asia is the active data frame.

data source

The data referenced by a layer or a layer file in ArcMap or ArcCatalog. Examples of data sources are a geodatabase feature class, a shapefile, and an image.

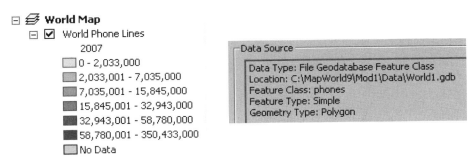

The data source for this World Phone Lines layer is the geodatabase feature class "phones" found in the World1 geodatabase (World1.mdb). The geodatabase is located in the C:\MapWorld\Mod1\ Data folder.

data view

A view in ArcMap for exploring, displaying, and querying geographic data. This view hides map elements such as titles, north arrows, and scale bars. Compare layout view.

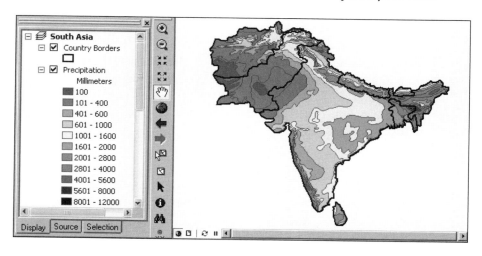

decimal degrees

Degrees of latitude and longitude expressed in decimals instead of minutes and seconds. Minutes and seconds are converted into a decimal using the mathematical formula below. In a GIS, using decimal degrees is more efficient than using minutes and seconds because it makes digital storage of coordinates easier and computations faster.

Decimal degrees = degrees + (minutes/60) + (seconds/3,600)

73° 59' 15" longitude = 73.9875 decimal degrees

feature

A geographic object on a map represented by a point, a line, or a polygon.

- A point feature is a map object that has no length or width, such as a tree on a neighborhood map or a city on a world map.
- A line feature is a one-dimensional map object such as a river or a street.
- A polygon feature is a two-dimensional map object such as a lake, a city, or a continent.

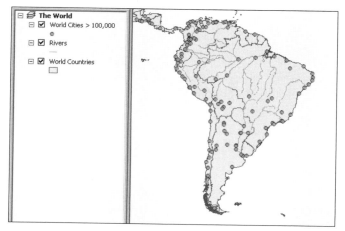

feature class

A collection of geographic features with the same geometry type (point, line, or polygon), the same attributes, and the same spatial reference (coordinate system and map projection).

field

The column in a table that contains the values (information) for a single attribute of each geographic feature in a GIS layer.

ObjectID	Shape ^	City Name	Country Name	Population	CAPITAL
0	Point	Guatemala	Guatemala	1,400,000	Y
1	Point	Tegucigalpa	Honduras	551,606	Y
2	Point	San Salvador	El Salvador	920,000	Y
3	Point	Managua	Nicaragua	682,000	Y
4	Point	San Jose	Costa Rica	670,000	Y
5	Point	Belmopan	Belize	4,500	Y
6	Point	Panama	Panama	625,000	Y
7	Point	San Jose	US	629,400	N

In this table, the City Name field contains the name for each city in this layer. The Population field contains the population value for each city.

field name

The column heading in an attribute table. Because field names are often abbreviated, ArcGIS allows you to create an alternative name, or alias, that can be more descriptive. In the graphic above, City Name and Country Name are aliases for fields named "NAME" and "COUNTRY."

Find button

An ArcMap button used for locating one or more map features that have a particular attribute value.

folder connection

A shortcut that allows you to navigate to a folder without having to enter the entire path.

geodatabase

A database used to organize and store geographic data in ArcGIS.

georeference

To assign coordinates from a reference system, such as latitude/longitude, to the page coordinates of an image or map.

graduated color map

A map that uses a range of colors to show a sequence of numeric values. For example, on a population density map, the more people per square kilometer, the darker the color.

graph

A graphic representation of tabular data.

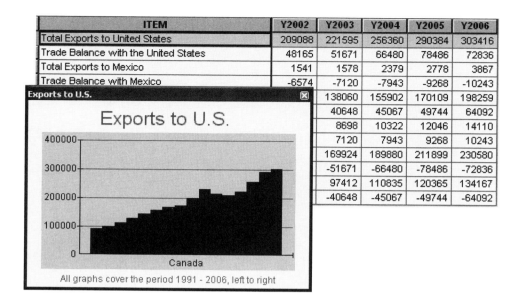

ITEM	Y2002	Y2003	Y2004	Y2005	Y2006
Total Exports to United States	209088	221595	256360	290384	303416
Trade Balance with the United States	48165	51671	66480	78486	72836
Total Exports to Mexico	1541	1578	2379	2778	3867
Trade Balance with Mexico	-6574	-7120	-7943	-9268	-10243
		138060	155902	170109	198259
		40648	45067	49744	64092
		8698	10322	12046	14110
		7120	7943	9268	10243
		169924	189880	211899	230580
		-51671	-66480	-78486	-72836
		97412	110835	120365	134167
		-40648	-45067	-49744	-64092

Identify tool

An ArcMap tool used to display the attributes of features in the map.

image

A graphic representation of data such as a scanned picture or a satellite photograph.

join

An operation that appends the fields of one table to those of another through an attribute field common to both tables. A join is usually used to attach more attributes to the attribute table of a map layer so that these attributes can be mapped. For example, you could join a country table with population data to a country layer attribute table. Compare relate.

label

Text placed next to a geographic feature on a map to describe or identify it. Feature labels usually come from an attribute field in the attribute table.

layer

A set of geographic features of the same type along with the associated attribute table, or an image. Examples of layers are "Major Cities," "Countries," and "Satellite Image." A layer references a specific data source such as a geodatabase feature class or image. Layers have properties, such as a layer name, symbology, and label placement. They can be stored in map documents (.mxd) or saved individually as layer files (.lyr). See also data source.

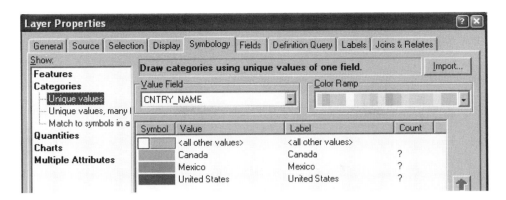

A layer has many properties, including Symbology properties. Some of the properties for the NAFTA Countries layer are pictured here.

layer, turn on

Turning on a layer allows the layer to be displayed in the map. In ArcMap, a layer is turned on by placing a check mark in the box next to the layer name in the table of contents.

layer file

In ArcGIS, a file with a .lyr extension that stores the path to a data source and other layer properties, including symbology.

layout

In ArcMap, an on-screen presentation document that can include maps, graphs, tables, text, and images.

layout view

A view in ArcMap in which geographic data and map elements, such as titles, legends, and scale bars, are placed and arranged for printing.

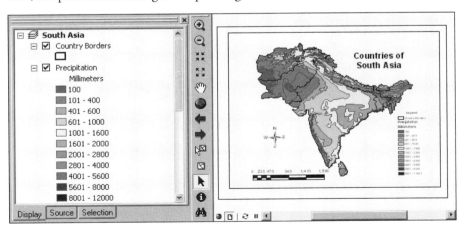

legend

A list identifying what each symbol represents on the map.

line

See feature.

Magnifier window

A window in ArcMap data view that shows a zoomed-in view of a small area of the main map. Moving the Magnifier window around does not change the extent of the map underneath.

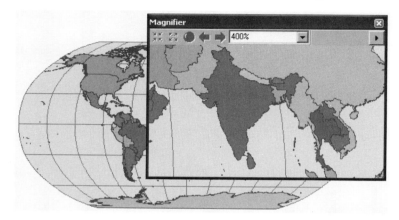

map document

In ArcMap, the file that contains one or more data frames and the associated layers, tables, graphs, and reports. Map document files have a .mxd extension.

map projection

A method by which the curved surface of the earth is portrayed on a flat map. Every map projection distorts distance, area, shape, direction, or some combination thereof. Map projections are made using complex mathematical formulas that are part of ArcGIS software's automatic functions.

MapTip

In ArcMap, a pop-up label for a map feature that is displayed when the mouse is paused over that feature. The label comes from a field in the layer attribute table.

Measure tool

An ArcMap tool used to measure distance on a map.

metadata

Information about the content, quality, condition, and other characteristics of data. Metadata may include a brief description of the data and its purpose, the names of the authors or compilers of the data, the date it was collected or created, the meaning of attribute fields, its scale, and its spatial reference (coordinate system and map projection).

pan

To move your map up, down, or sideways without changing the viewing scale.

point

See feature.

polygon

See feature.

projection

See map projection.

record

A row in an attribute table that contains all of the attribute values for a single feature.

ObjectID	Shape *	City Name	Country Name	Population
0	Point	Guatemala	Guatemala	1,400,000
1	Point	Tegucigalpa	Honduras	551,606
2	Point	San Salvador	El Salvador	920,000
3	Point	Managua	Nicaragua	682,000
4	Point	San Jose	Costa Rica	670,000
5	Point	Belmopan	Belize	4,500
6	Point	Panama	Panama	625,000
7	Point	San Jose	US	629,400

This table has seven records. The fourth record is highlighted. It contains all of the attributes for the point feature representing the city of Managua, Nicaragua.

relate

An operation that establishes a temporary connection between records in two tables using a field common to both. Unlike a join operation, a relate does not append the fields of one table to the other. A relate is usually used to associate more records and their attributes to the attribute table of a map layer. For example, you could relate a table listing large cities to a layer attribute table of countries, or you could join a world cities table to a country layer attribute table. Compare join.

scale

The relationship between a distance or area on a map and the corresponding distance or area on the ground, commonly expressed as a fraction or ratio. A map scale of 1/100,000 or 1:100,000 means that one unit of measure (e.g., one inch) on the map equals 100,000 of the same units on the earth.

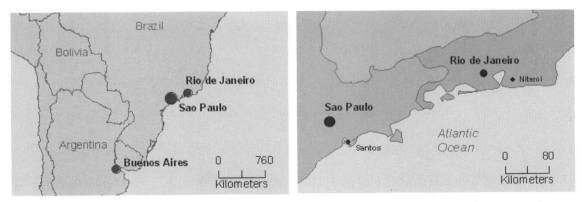

The map on the left has a map scale of 1:80,000,000. The map on the right, which is zoomed in, has a map scale of 1:6,000,000.

selected feature

A geographic feature that is chosen and put into a subset so that various functions can be performed on the feature. In ArcMap, a feature can be selected by clicking it on the map with the Select Features tool or highlighting an attribute in a table. When a geographic feature is selected, it is outlined in blue on the map. Its corresponding record in the attribute table is highlighted in blue.

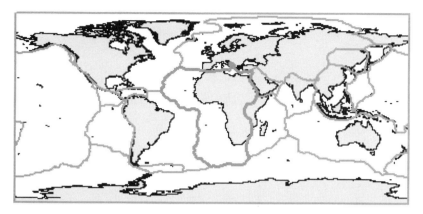

The Africa plate is selected in this map of the earth's tectonic plates.

shapefile (.shp)

A data storage format for storing the location, shape, and attribute information of geographic features. A shapefile is stored in a set of related files and contains one feature class.

source data

See data source.

sort ascending

To arrange an attribute table's rows in order from the lowest values to the highest values in a field. For example, number values would be ordered from 1 to 100, and alphabetical values would be ordered from A to Z.

sort descending

To arrange an attribute table's rows in order from the highest to the lowest values in a field. For example, number values would be ordered from 100 to 1, and alphabetical values would be ordered from Z to A.

Symbol Selector

The dialog in ArcMap for selecting symbols and changing their color, size, outline, or other properties.

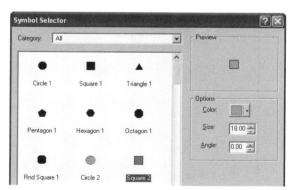

table of contents

A list of data frames and layers on a map that may also show how the data is symbolized.

```
☐ ≋ The World
    ☐ ☑ World Cities > 100,000
            ●
    ☐ ☑ Rivers

    ☐ ☑ World Countries
            ☐
☐ ≋ Standard of Living Indicators
    ⊞ ☐ Infant Mortality Rate
    ☐ ☑ Life Expectancy
            Years
            ☐ 32 - 46
            ☐ 47 - 57
            ☐ 58 - 67
            ▨ 68- 75
            ■ 76 - 84
            ▨ No Data
    ⊞ ☐ Literacy Rate
```

toolbar

A set of commands that allows you to carry out related tasks. The Main Menu toolbar in ArcMap has a set of menu commands; other toolbars typically have buttons. Toolbars can float on the desktop in their own window or may be docked at the top, bottom, or sides of the main window.

vertex

One of the points that defines a line or polygon feature.

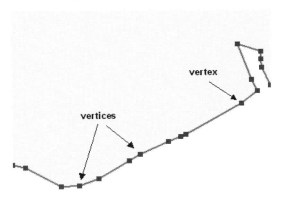

zoom

To display a larger or smaller extent of a GIS map or image.

Data sources

Module 1 - \OurWorld2\Mod1\Data
Data sources include:
\World1.gdb\Cities_GR_100K feature class from ESRI Data & Maps 2006, courtesy of ArcWorld.

\World1.gdb\Cities_GR_5M feature class from ESRI Data & Maps 2006, courtesy of ArcWorld.

\World1.gdb\cntry07_demog combined feature class from ESRI Data & Maps 2007, courtesy of ArcWorld; the US Census Bureau, International Division; and the CIA World Factbook.

\World1.gdb\cntry07_econ combined feature class from ESRI Data & Maps 2007, courtesy of ArcWorld and the CIA World Factbook.

\World1.gdb\geogrid feature class from ESRI Data & Maps 2006.

\World1.gdb\lakes feature class from ESRI Data & Maps 2006, courtesy of ArcWorld.

\World1.gdb\rivers feature class from ESRI Data & Maps 2006, courtesy of ArcWorld.

Module 2 - \OurWorld2\Mod2\Data
Data sources include:
\Images\earth_wsi.sid from ESRI Data & Maps 2004, satellite image courtesy of WorldSat, Inc.

\EastAsia.gdb\popdens05 feature class from Center for International Earth Science Information Network (CIESIN), The Earth Institute, Columbia UniversityArchives Manager, and NASA Socioeconomic Data and Applications Center (SEDAC).

\World2.gdb\cntry07 feature class from ESRI Data & Maps 2007, courtesy of ArcWorld

\World2.gdb\continents feature class from ESRI Data & Maps 2004, courtesy of ArcWorld Supplement.

\World2.gdb\earthquakes feature class from U.S. Geographical Survey National Earthquake Information Center.

\World2.gdb\faults feature class from ArcAtlas: Our Earth, 1996, courtesy of Data+ and ESRI.

\World2.gdb\features feature class from Mapping Our World: GIS Lessons for Educators ArcGIS Desktop Edition, ESRI Press, 2005. Original data courtesy of National Geographic Maps.

\World2.gdb\major_cities feature class from ESRI Data & Map 2004, courtesy of DMTI Spatial Inc.

\World2.gdb\plates feature class from ArcAtlas: Our Earth, 1996, courtesy Data+ and ESRI, and edited by Roger Palmer based on ocean bathymetry from WSI_Earth.sid in comparison with USGS plate lines.

\World2.gdb\plates_line feature class from ArcAtlas: Our Earth, 1996, courtesy Data+ and ESRI, and edited by Roger Palmer based on ocean bathymetry from WSI_Earth.sid in comparison with USGS plate lines

\World2.gdb\volcanoes feature class from ArcAtlas: Our Earth, 1996, courtesy Data+ and ESRI; created from plate lines data edited by Roger Palmer based on ocean bathymetry from WSI_Earth.sid in comparison with USGS plate lines.

\World2.gdb\world30 feature class from ESRI Data & Maps 2004.

Module 3 - \OurWorld2\Mod3\Data
Data sources include:

\SouthAsia.gdb\agriculture feature class from *Mapping Our World: GIS Lessons for Educators ArcGIS Desktop Edition,* ESRI Press, 2005; original data courtesy of Data+ and ESRI, ArcAtlas: Our Earth.

\SouthAsia.gdb\cities_sa feature class from ESRI Data & Maps, 2004.

\SouthAsia.gdb\cntry_sa feature class from *Mapping Our World: GIS Lessons for Educators ArcGIS Desktop Edition,* ESRI Press, 2005; original data from ESRI Data & Maps 2001, courtesy of ArcWorld.

\SouthAsia.gdb\pop_dens feature class from *Mapping Our World: GIS Lessons for Educators ArcGIS Desktop Edition,* ESRI Press, 2005; original data courtesy of Data+ and ESRI, ArcAtlas: Our Earth.

\SouthAsia.gdb\precip_sa feature class from *Mapping Our World: GIS Lessons for Educators ArcGIS Desktop Edition,* ESRI Press, 2005; original data courtesy of Data+ and ESRI, ArcAtlas: Our Earth.

\SouthAsia.gdb\relief_sa feature class from *Mapping Our World: GIS Lessons for Educators ArcGIS Desktop Edition,* ESRI Press, 2005; original data courtesy of Data+ and ESRI, ArcAtlas: Our Earth.

\SouthAsia.gdb\rivers_sa feature class from *Mapping Our World: GIS Lessons for Educators ArcGIS Desktop Edition,* ESRI Press, 2005; original data courtesy of Data+ and ESRI, ArcAtlas: Our Earth.

\World3.gdb\climate feature class from *Mapping Our World: GIS Lessons for Educators ArcGIS Desktop Edition,* ESRI Press 2005; original data courtesy of National Geographic Maps.

\World3.gdb\cntry07_demog feature class from ESRI Data & Maps 2007 Update, courtesy of ArcWorld Supplement; US Census Bureau, International Division; and the CIA World Factbook.

\World3.gdb\geogrid feature class from ESRI Data & Maps 2004.

\World3.gdb\lakes feature class from ESRI Data & Maps 2004, courtesy of ArcWorld.

\World3.gdb\rivers feature class from ESRI Data & Maps 2004, courtesy of ArcWorld.

\World3.gdb\temp_city combined feature class from Worldclimate.com and ESRI Data and Maps 2004, author added Celsius in addition to Fahrenheit to the data.

\World3.gdb\world30 feature class from ESRI Data & Maps 2004.

\WorldPopDensity05.gdb\WorldPopDensity05 feature class from Center for International Earth Science Information Network (CIESIN) the Earth Institute, Columbia University

Archives Manager, and NASA Socioeconomic Data and Applications Center (SEDAC).

Module 4 - \OurWorld2\Mod4\Data
Data sources include:

\World4.gdb\TopTen\city100 feature class derived from Tertius Chandler's book, *Four Thousand Years of Urban Growth: An Historical Census,* Edwin Mellen Press,1987.

\World4.gdb\TopTen\city1000 feature class derived from Tertius Chandler's book, *Four Thousand Years of Urban Growth: An Historical Census,* Edwin Mellen Press,1987.

\World4.gdb\TopTen\city1500 feature class derived from Tertius Chandler's book, *Four Thousand Years of Urban Growth: An Historical Census,* Edwin Mellen Press,1987.

\World4.gdb\TopTen\city1800 feature class derived from Tertius Chandler's book, *Four Thousand Years of Urban Growth: An Historical Census,* Edwin Mellen Press,1987.

\World4.gdb\TopTen\city1900 feature class derived from Tertius Chandler's book, *Four Thousand Years of Urban Growth: An Historical Census,* Edwin Mellen Press,1987.

\World4.gdb\TopTen\city1950 feature class derived from Tertius Chandler's book, *Four Thousand Years of Urban Growth: An Historical Census,* Edwin Mellen Press,1987.

\World4.gdb\TopTen\city2000 combined feature class from ESRI Data & Maps 2006 (cities) and the United Nations Population Division (population).

\World4.gdb\TopTen\city2005 combined feature class from ESRI Data & Maps 2006 (cities) and the United Nations Population Division (population).

\\World4.gdb\cntry07_econ combined feature class from ESRI Data & Maps 2007 courtesy of ArcWorld and the CIA World Factbook.

\World4.gdb\cntry07_social combined feature class from ESRI Data & Maps 2007, courtesy of ArcWorld; the U.S. Census Bureau, International Division; and the CIA World Factbook.

\World4.gdb\continents feature class from ESRI Data & Maps 2004, courtesy of ArcWorld Supplement

\World4.gdb\geogrid feature class from ESRI Data & Maps 2004.

\World4.gdb\lakes feature class from ESRI Data & Maps 2004, courtesy of ArcWorld.

\World4.gdb\rivers feature class from ESRI Data & Maps 2004, courtesy of ArcWorld.

\World4.gdb\top30cities combined feature class from ESRI Data & Maps 2006 (cities) and United Nations Population Division (population).

\World4.gdb\vital_statistics combined feature class from ESRI Data & Maps 2007, courtesy of ArcWorld; the U.S. Census Bureau, International Division; and the CIA World Factbook.

\World4.gdb\world30 feature class from ESRI Data & Maps 2004.

Module 5 - \OurWorld2\Mod5\Data
Data sources include:

\Images\earth_wsi.sid courtesy of WorldSat International, Inc.

\Images\middle_east.tif courtesy of USGS, 2001, Global GIS Database: Digital Atlas of Africa.

\MiddleEast.gdb\Boundaries\ap_cntry feature class from ESRI Data & Maps 2000, courtesy of ArcWorld Supplement.

\MiddleEast.gdb\ Boundaries\ap_line feature class from ESRI Data & Maps 2004, courtesy of ArcWorld Supplement.

\MiddleEast.gdb\ Boundaries\arab_pen feature class from ESRI Data & Maps 2004, courtesy of ArcWorld Supplement.

\MiddleEast.gdb\ Boundaries\ Arabian_Peninsula___namesAnno feature class derived from ESRI Data & Maps, 2004, courtesy of ArcWorld Supplement.

\MiddleEast.gdb\ Boundaries\neighbors feature class from ESRI Data & Maps 2004, courtesy of ArcWorld Supplement.

\MiddleEast.gdb\ Boundaries\yemen1 feature class digitized from paper maps: *British-Yemeni Society Journal,* 2000.

\MiddleEast.gdb\ Boundaries\yemen2 feature class digitized from paper maps: *British-Yemeni Society Journal,* 2000.

\MiddleEast.gdb\ Boundaries\yemen3 feature class digitized from paper maps: *British-Yemeni Society Journal,* 2000.

\MiddleEast.gdb\Cities\city_town feature class from USGS; Digital Atlas of the Middle East (NIMA data).

\MiddleEast.gdb\Cities\major_cities feature class from USGS, 2001, Global GIS Database: Digital Atlas of Africa; from file gaz_cities.shp

\MiddleEast.gdb\agriculture feature class from ArcAtlas: Our Earth; Data+ and ESRI, 1996.

\MiddleEast.gdb\eco_zone feature class from ESRI Data & Maps 2004, courtesy of World Wildlife Fund.

\MiddleEast.gdb\pop_density feature class from ArcAtlas: Our Earth, courtesy Data+ and ESRI, 1996.

\MiddleEast.gdb\precipitation feature class from USGS, 2001, Global GIS Database: Digital Atlas of Africa (NIMA data).

\MiddleEast.gdb\roads feature class from USGS, 2001, Global GIS Database: Digital Atlas of Africa (NIMA data).

\MiddleEast.gdb\springs feature class from USGS, 2001, Global GIS Database: Digital Atlas of Africa (NIMA data).

\MiddleEast.gdb\streams feature class from USGS, 2001, Global GIS Database: Digital Atlas of Africa (NIMA data).

\MiddleEast.gdb\temperature feature class from USGS, 2001, Global GIS Database: Digital Atlas of Africa (NIMA data).

\MiddleEast.gdb\waterbodies feature class from USGS, 2001, Global GIS Database: Digital Atlas of Africa (NIMA data).

\World5.gdb\cntry07_demog combined feature class from ESRI Data & Maps 2007, courtesy of ArcWorld; the US Census Bureau, International Division; and the CIA World Factbook.

\World5.gdb\cntry92 feature class from ESRI Data & Maps 2004, courtesy of ArcWorld.

\World5.gdb\lakes feature class from ESRI Data & Maps, courtesy of ArcWorld.

\World5.gdb\Language feature class from Mapping Our World: GIS Lessons for Educators ArcGIS Desktop Edition, ESRI Press, 2005; original data courtesy of National Geographic Maps.

\World5.gdb\oil_gas feature class from ArcAtlas: Our Earth; Data+ and ESRI, 1996.

\World5.gdb\Religion ESRI feature class from Mapping Our World: GIS Lessons for Educators ArcGIS Desktop Edition, ESRI Press, 2005; original data courtesy of National Geographic Maps.

\World5.gdb\rivers feature class from ESRI Data & Maps 2004, courtesy of ArcWorld.

\World5.gdb\world30 feature class from ESRI Data & Maps 2004.

Module 6 - \OurWorld2\Mod6\Data\
Data sources include:

\NAFTA.gdb\NAFTA_countries combined feature class from ESRI Data & Maps 2004 and U.S. Census.

\NAFTA.gdb\NAFTA_Trading_Statistics combined feature class from ESRI Data & Maps 2004 and U.S. Census.

\World6.gdb\cntry04_energy combined feature class from ESRI Data & Maps 2004 and the United States Energy Information Administration.

\World6.gdb\cntry04_social combined feature class from ESRI Data & Maps 2007 and the CIA World Factbook.

\World6.gdb\cntry07_demog combined feature class from ESRI Data & Maps 2007, courtesy of ArcWorld; the U.S. Census Bureau, International Division; and the CIA World Factbook.

\World6.gdb\cntry07_econ combined feature class from ESRI Data & Maps 2007, courtesy of ArcWorld and the CIA World Factbook.

Module 7 - \OurWorld2\Mod7\Data
Data sources incude:

\Images\amery.tif created by USGS using Digital Elevation Model data from USGS TerraWeb; http://TerraWeb.wr.usgs.gov 2001.

\Images\ellswrth.tif created by USGS using Digital Elevation Model data from USGS TerraWeb; http://TerraWeb.wr.usgs.gov 2001.

\Images\larsen_breakup.tif recorded by NASA's MODIS satellite sensor: NASA/Goddard Space Flight Center.

\Images\mcmurdo.tif created by USGS using Digital Elevation Model data from USGS TerraWeb; http://TerraWeb.wr.usgs.gov 2001.

\Images\min120m.jpg created using GTOPO30 and Spatial Analyst 2001.

\Images\mitch2sat.tif from USGS Digital Atlas of Central America, NOAA, NASA.

\Images\mitch3sat.tif from USGS Digital Atlas of Central America, NOAA, NASA.

\Images\mitch4sat.tif from USGS Digital Atlas of Central America, NOAA, NASA.

\Images\mitch5sat.tif from USGS Digital Atlas of Central America, NOAA, NASA.

\Images\plus50m.jpg created using GTOPO30 and Spatial Analyst 2002.

\Images\plus5m.jpg created using GTOPO30 and Spatial Analyst 2002.

\Images\plus73m.jpg created using GTOPO30 and Spatial Analyst 2002.

\Images\sealevel.jpg created using GTOPO30 and Spatial Analyst 2002.

\Images\southpole2.tif from NOAA Corps Collection.

\Antarctica.gdb\S_Pole_pts feature class digitized by ESRI from paper map, "Political and Physical Map of the World," courtesy National Geographic Society.

\CentralAmerica.gdb\agr_use feature class from ArcAtlas Our Earth, courtesy of Data+ and ESRI.

\CentralAmerica.gdb\ca_airportsfeature class from U.S. Geological Survey, Global GIS Database: Digital Atlas of Central America.

\CentralAmerica.gdb\ca_capitals feature class from ESRI Data & Maps 2000, courtesy of ArcWorld.

\CentralAmerica.gdb\ca_countries feature class from ESRI Data & Maps 2000, courtesy of ArcWorld Supplement.

\CentralAmerica.gdb\ca_drain data origin Digital Chart of the World gazetteer.

\CentralAmerica.gdb\ca_railroads feature class from U.S. Geological Survey, Global GIS Database: Digital Atlas of Central America.

\CentralAmerica.gdb\ca_roads feature class from U.S. Geological Survey, Global GIS Database: Digital Atlas of Central America.

\CentralAmerica.gdb\ca_utility feature class from U.S. Geological Survey, Global GIS Database: Digital Atlas of Central America.

\CentralAmerica.gdb\coastal data origin Digital Chart of the World gazetteer.

\CentralAmerica.gdb\landforms data origin Digital Chart of the World gazetteer.

\CentralAmerica.gdb\mitch2 courtesy of U.S. Geological Survey Digital Atlas of Central America, NOAA.

\CentralAmerica.gdb\mitch3 courtesy of U.S. Geological Survey Digital Atlas of Central America, NOAA, NASA.

\CentralAmerica.gdb\mitch4 courtesy of U.S. Geological Survey Digital Atlas of Central America, NOAA, NASA.

\CentralAmerica.gdb\mitch5 courtesy of U.S. Geological Survey Digital Atlas of Central America, NOAA, and NASA.

\CentralAmerica.gdb\mitch_pre feature class from U.S. Geological Survey, Digital Atlas of Central America & NOAA.

\CentralAmerica.gdb\pop_plc feature class from U.S. Geolgocial Survey, Digital Atlas of Central America & NOAA.

\CentralAmerica.gdb\precip feature class from ArcAtlas: Our Earth, 1996. courtesy of Data+ and ESRI.

\CentralAmerica.gdb\rain3 feature class from USGS Digital Atlas of Central America & NOAA.

\CentralAmerica.gdb\rain4 feature class from USGS Digital Atlas of Central America & NOAA.

\CentralAmerica.gdb\rain5 feature class from USGS Digital Atlas of Central America & NOAA.

\World7.gdb\airports feature class from from ArcAtlas: Our Earth, 1996, courtesy of Data+ and ESRI.

\World7.gdb\cntry07_demog combined feature class from ESRI Data & Maps 2007, courtesy of ArcWorld and the CIA World Factbook.

\World7.gdb\continents feature class from ESRI Data & Maps 2004, courtesy ArcWorld Supplement.

\World7.gdb\energy feature class from ArcAtlas: Our Earth, 1996, courtesy of Data+ and ESRI.

\World7.gdb\geogrid feature class from ESRI Data & Maps 2006.

\World7.gdb\lakes feature calss from ESRI Data & Maps 2006, courtesy of ArcWorld.

\World7.gdb\latlong feature class from ESRI Data & Maps 2004.

\World7.gdb\major_cities feature class from ESRI Data & Maps 2000, courtesy of ArcWorld.

\World7.gdb\manufact_plc feature class from ArcAtlas: Our Earth, 1996, courtesy of Data+ and ESRI.

\World7.gdb\mineral_res feature class from ArcAtlas: Our Earth, 1996, courtesy of Data+ and ESRI.

\World7.gdb\pipelines feature class from ArcAtlas: Our Earth, 1996, courtesy of Data+ and ESRI.

\World7.gdb\rivers feature class from ESRI Data & Maps 2006, courtesy of ArcWorld.

\World7.gdb\roads_rail feature class from ArcAtlas: Our Earth, 1996, courtesy of Data+ and ESRI.

\World7.gdb\us_cities feature class from ESRI Data & Maps 2006, courtesy of ArcWorld.

\World7.gdb\w_cities feature class from ESRI Data & Maps 2000, courtesy of ArcWorld.

\World7.gdb\world30 feature class from ESRI Data & Maps 2006.

\World7.gdb\WWF_Eco feature class from ESRI Data & Maps 2004, courtesy of World Wildlife Fund.